3D Draughting Using AutoCAD

Robert McFarlane

*Senior Lecturer, Department of Integrated
Engineering, Motherwell College*

A member of the Hodder Headline Group
LONDON • SYDNEY • AUCKLAND
Copublished in North, Central and South America
by John Wiley & Sons, Inc., New York • Toronto

First published in Great Britain in 1997 by Arnold,
a member of the Hodder Headline Group,
338 Euston Road, London NW1 3BH

Copublished in North, Central and South America by
John Wiley & Sons, Inc., 605 Third Avenue
New York, NY 10158-0012

Whilst the advice and information in this book is believed to be true and
accurate at the date of going to press, neither the author nor the publisher
can accept any legal responsibility or liability for any errors or omissions
that may be made.

British Library Cataloguing in Publication Data
A catalogue record for this book is available from the British Library

Library of Congress Cataloging-in-Publication Data
A catalog record for this book is available from the Library of Congress

ISBN 0 340 67782 1
ISBN 0 470 23732 5 (Wiley)

Produced and typeset in 10/12 pt Garamond by Gray Publishing, Tunbridge Wells
Printed and bound in Great Britain by JW Arrowsmith Ltd, Bristol

Contents

1

Introduction

AutoCAD is presently the most widely used PC computer-aided draughting program. It is very user-friendly, and will allow users to produce drawings after a few hours of tuition.

This package will introduce the user to three-dimensional (3D) draughting, and consists of:
- a set of notes consisting of worked examples and drawings
- a two-disk set containing **all** the worked examples.

USING THE PACKAGE

Each chapter is a self-contained module which will introduce the user to a new concept. The notes are supplemented with example exercises.

There are several conventions which the user must be aware of, these being:
1. Keyboard entries and menu item selections will be in **boldface** type, for example
   (a) select from the screen menu **DRAW**
                                   CIRCLE
   (b) select from the menu bar **Settings**
                                   Layer Control...
   (c) enter a radius **15**.
2. AutoCAD prompts will be given as: 'prompt:' followed by the program's message.
3. The return (or enter) key will be given as <R> or <RETURN>.
4. Users should have a reasonable working knowledge of the AutoCAD commands and how to use dialogue boxes.

COPYING THE DISKS

The two disks supplied contain all the files which will be used during the various exercises. All the files should be copied into a directory on the hard drive of your computer, and the following procedure will make this directory and copy all the files:
1. Start your system with the C:\> prompt active.
2. Insert disk 1 into the drive.
3. Type **A:\INSTALL<R>**
4. Follow the instructions as they appear on the screen.

All exercises and activities will be loaded from the created directory called 3DPACK. The user may want to have a blank formatted floppy available to save work, although all work can be saved to the created directory.

SYSTEM REQUIREMENTS

AutoCAD R12 should be installed on either a 386 or 486 machine and a mouse or tablet should be used as the pointing device. My system is an Elonex PC-433 and I used the Elonex two-button mouse.

2

Three-dimensional extruded drawings

Extruded drawings were one of the first excursions by AutoCAD into three-dimensional (3D) draughting. Extruded drawings are created from line, circle and arc entities by (as the name suggests) 'extruding' a shape upwards or downwards from a datum plane. The resultant model is not a 'real' 3D component and has been called two-and-a-half dimensional ($2^{1}/_{2}$D) draughting.

Extruded 3D drawings are a very good way to introduce new users to some of the commands which are used with 3D draughting. They also have their own terminology which is displayed in Fig. 2.1 and is:

- elevation – the plane from which the extrusion is drawn
- thickness – the height/depth of the extruded shape and:
  - (a) positive thickness, extrudes upwards
  - (b) negative thickness, extrudes downwards
- hide – displays a model with hidden line removal.

(a)

(b)

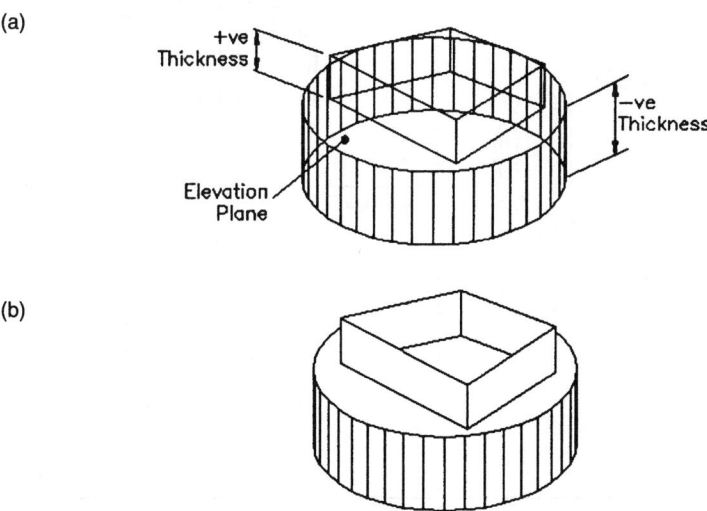

Fig. 2.1 Three-dimensional extrusion with (**a**) basic terminology and (**b**) with HIDE.

OPENING A DRAWING FILE

All drawing files are opened in the same manner, and the following sequence will open drawing EX2_1 from the 3DPACK directory:
1. Start AutoCAD
2. From the menu bar select **File**

                             Open ...

    prompt: Open Drawing dialogue box
    respond: (a) double left click on \
                  (b) double left click on 3DPACK
                  (c) pick drawing file EX2_1
                  (d) pick OK
3. The 3D extruded component will be displayed in red in three viewports as Fig. 2.2, with layer OUT (red) current.

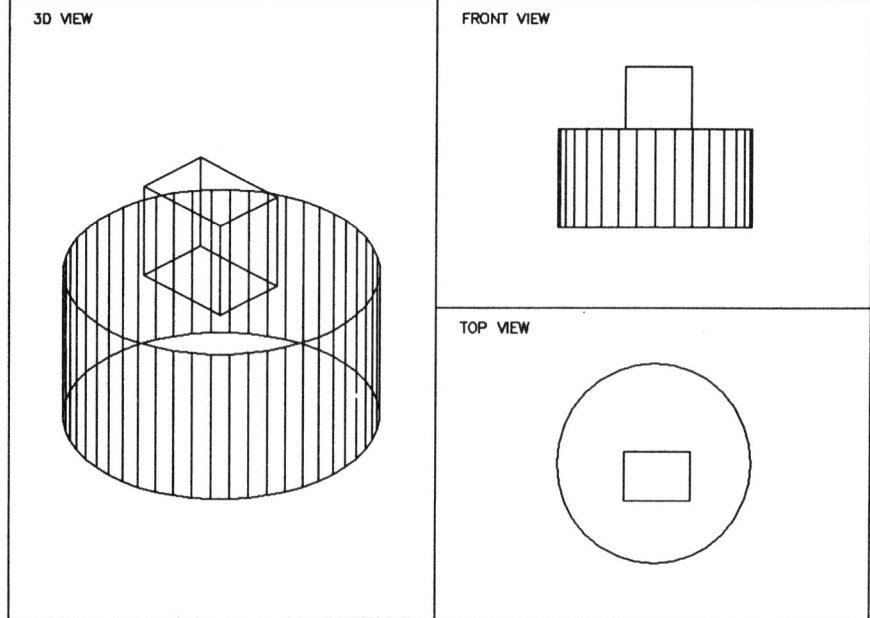

Fig. 2.2   Three-dimensional extruded example 1 – original model.

CHANGING COMPONENT COLOURS

1. Ensure the 3D viewport is active by moving the mouse and left-click in the required viewport – it probably is active?
2. From the screen menu select **EDIT–CHPROP**
   prompt: Select objects
   respond: **pick the cylindrical object then <R>**
   prompt: Change what property…
   respond: **pick Colour from the screen**
   prompt: New Colour
   respond: **pick blue from the screen**
   prompt: Change what property… – i.e. any more to be changed?
   respond: <RETURN> to end sequence.
3. The 'cylinder' should now be blue in all viewports.

ADDING ADDITIONAL ENTITIES TO THE MODEL

1. Activate the TOP viewport.
2. From the screen menu select **AutoCAD–SETTINGS–ELEV:**
   prompt: New current elevation <0.00>
   enter: **0 <R>**
   prompt: New current thickness <0.00>
   enter: **80 <R>**
3. Select from the screen menu **DRAW–CIRCLE–CEN,RAD**
   prompt: 3P/2P/TTR/<Center point>
   enter: **200,45 <R>**
   prompt: Diameter/<Radius>
   enter: **10 <R>**
4. A red 'cylinder' will be displayed in all viewports.
5. Using EDIT-CHPROP, change the colour of this last cylinder to green.
6. From the menu bar select **Construct**
                          Array
   prompt: Select objects
   respond: **pick the green cylinder then <R>**
   prompt: Rectangular or Polar array (R/P)
   enter: **P <R>**
   prompt: Center point of array
   respond: (a) **pick **** then CENtre** from the screen menu
           (b) **pick the blue circle then <R>**
   prompt: Number of items
   enter: **8 <R>**
   prompt: Angle to fill…
   enter: **360 <R>** – for a complete circle fill
   prompt: Rotate objects as they are copied
   enter: **Y <R>**

7. The component now has eight green cylinders added to it.
8. From the menu bar select **Settings**
                              Entity Modes...
   prompt: Entity Creation Modes dialogue box
   respond: (a) alter the Elevation to 130
             (b) alter the Thickness to 30
             (c) pick OK
9. Draw a CIRCLE, centre at 200,90 with a radius of 20, and change the colour of the circle to magenta.
10. Using the Entity Modes dialogue box:
    (a) set the elevation to 0.
    (b) set the thickness to −50.
11. Draw a circle, centre 200,100 with a radius of 50. Make this circle colour yellow.
12. Your screen display should now resemble Fig. 2.3 which has been plotted with HIDE.

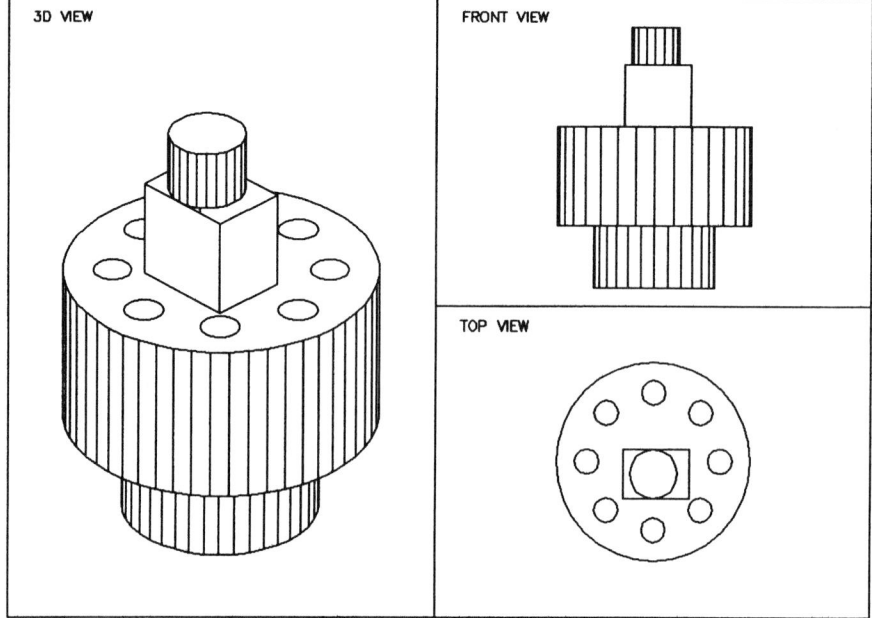

Fig. 2.3   Three-dimensional example 1 with added entities (plotted with HIDE).

HIDE, SHADE AND RENDER

All three-dimensional drawings can be difficult to 'see' due to the number of lines making up the component. AutoCAD has several viewing aids available, and these will now be investigated.

1. With the 3D viewport active, enter **HIDE** <R> and the extruded model will be displayed with **hidden line removal**.
2. At the command line enter **REGEN** <R> to return the model to its complete visualization.
3. From the screen menu select **DISPLAY**

   SHADE

   SHADE:

4. A nicely coloured extruded component will be displayed.
5. Return the model to its original display with REGEN.
6. From the menu bar select **Render**

   Render

   Note: it may be necessary to <RETURN> five times with the render command.
7. A very pleasing coloured image of the component will be obtained.
8. Return to the drawing screen with <RETURN>

Notes

1. Hide, Shade and Render can be used in any viewport as long as it is active.
2. The three commands can be easily entered at the command line i.e. HIDE, SHADE, RENDER.

SAVING YOUR WORK

Our first extruded exercise is now complete, and you may want to save your work. The drawing can be saved either to the 3DPACK directory or to a floppy disk – it's your decision. If you save to the directory, make sure that you give the drawing a different name from EX2_1. The following is my preference:

1. Insert a blank formatted disk into the A drive.
2. Select from the menu bar **File–Save As...**
3. From the Save As dialogue box, select **Type It**
4. Enter the name **A:EXTR1** <R> at the command line.

THE VPOINT COMMAND

The viewpoint (VPOINT) command allows 3D models to be viewed from different angles and from above or from below. We will demonstrate the command with a new example, so:

1. Open the drawing **EX2_2** from the 3DPACK directory.
2. Ensure the top left viewport is active.
3. From the screen menu select **DISPLAY**
                                   VPOINT
                                   rotate
   prompt: Enter angle in *XY* plane from *X* axis
   enter: **80 <R>**
   prompt: Enter angle from *XY* plane
   enter: **30 <R>**
4. With the VPOINT–rotate option, enter the following values in the specified viewport:

	Angle in XY plane	Angle from XY plane
top right	80	−30
bottom right	315	60
bottom left	315	−60

5. In each viewport use the HIDE command, and the result should be as Fig. 2.4.
6. Save the alterations if required. Try SHADE, RENDER?

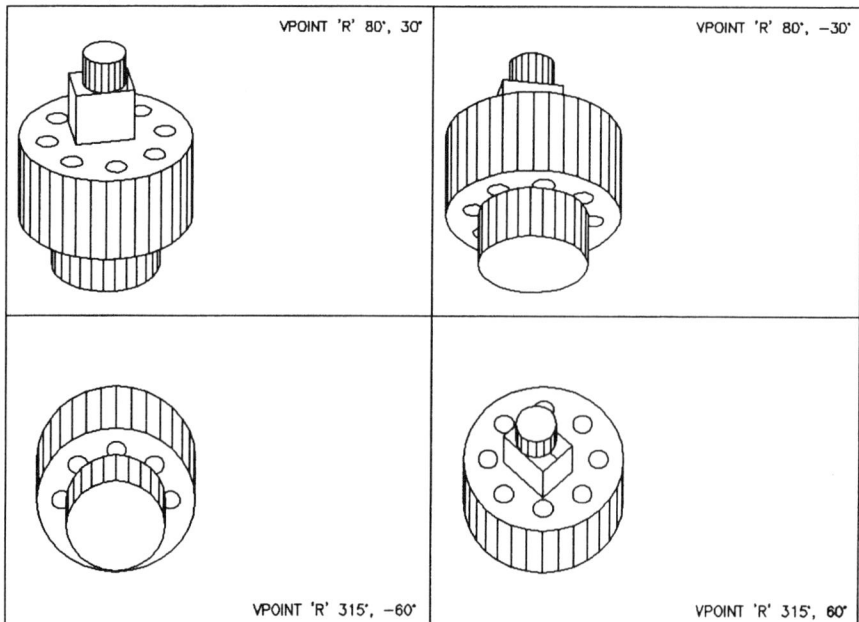

VPOINT 'R' 80°, 30°

VPOINT 'R' 80°, −30°

VPOINT 'R' 315°, −60°

VPOINT 'R' 315°, 60°

Fig. 2.4   Three-dimensional extruded example 2 – after the VPOINT 'R' options.

THE VPOINT–ROTATE OPTION

The VPOINT command has several options, and we have only investigated one of these
– the rotate option. This has two prompts:
- *prompt 1*: angle in *XY*-plane. This is the users 'standpoint' on the *XY*-plane looking
  at the component. It can be any value between 0° and 360°, the 0 value looking at
  the component from the right side
- *prompt 2*: angle from *XY*-plane. This is the users 'eye angle' from the *XY*-plane, i.e.
  looking down/up at the component and
  (a) positive angle is looking down from above
  (b) negative angle is looking up from below.

SUMMARY

1. Three-dimensional extruded drawings are created with ELEV and THICKNESS.
2. Extruded models are created from line, circle and arc entities.
3. An extruded 3D model is not a real 3D model, as there are no top or bottom surfaces.
   The extruded model consists of 'sides'.
4. The ELEVATION command can be activated by:
   (a) SETTINGS from the screen menu
   (b) selecting the Entity Modes dialogue box from the Settings menu bar
   (c) entering ELEV at the command line.
5. Useful commands are:

HIDE	displays the model with hidden line removal in the active viewport.
REDRAW	redraws the active viewport
REDRAWALL	redraws all viewports.
REGEN	regenerates the active viewport, restoring the model after the HIDE command has been used.
REGENALL	regenerates all viewports.
SHADE	displays the active viewport model as a coloured drawing.
RENDER	displays the active viewport as a coloured rendered image.

ACTIVITY

1. Open the ACT_1 drawing from the 3DPACK directory – Fig. 2.5.
2. The extruded components consist of:
   (a) a red rectangular base (100 × 80) with ELEV 0 and THICK 30
   (b) four blue square pillars (10 × 10) with ELEV 30 and THICK 80.
3. Complete the following:
   (a) set ELEV to 0 and THICK to 110, and draw a circle (green) with centre at the point −10,−10 and radius 10
   (b) rectangular array the green circle by two rows and two columns. The row distance is 100 and the column distance is 120
   (c) set ELEV to 110 and THICK to 20
   (d) use LINE to draw a rectangle using the green circle centres as the vertices of the rectangle.
4. The result should be as shown in Fig. 2.6.
5. In the 3D viewport try the HIDE, SHADE and RENDER commands.
6. Use the VPOINT-rotate option to change the viewpoint of the model. Figure 2.7 gives some ideas.

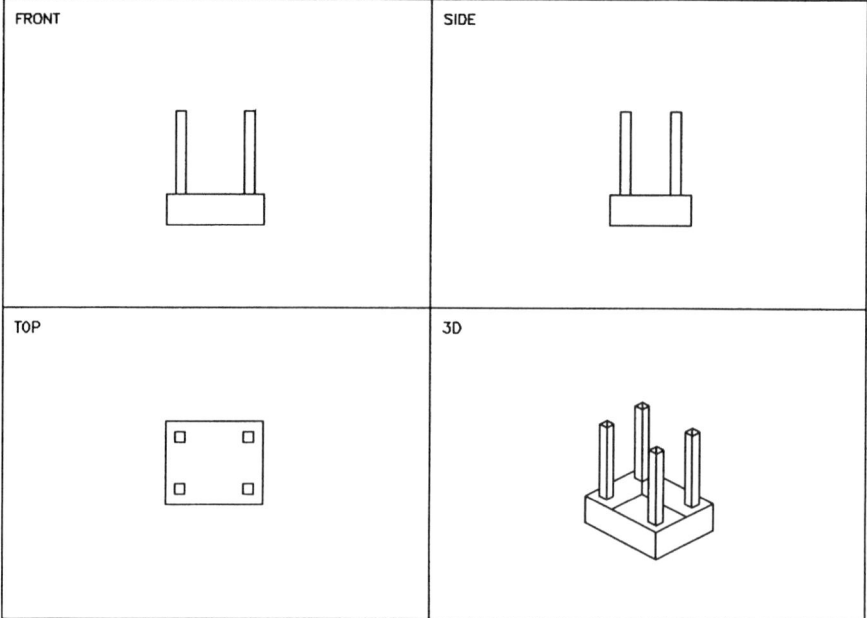

Fig. 2.5   Activity 1 – original component.

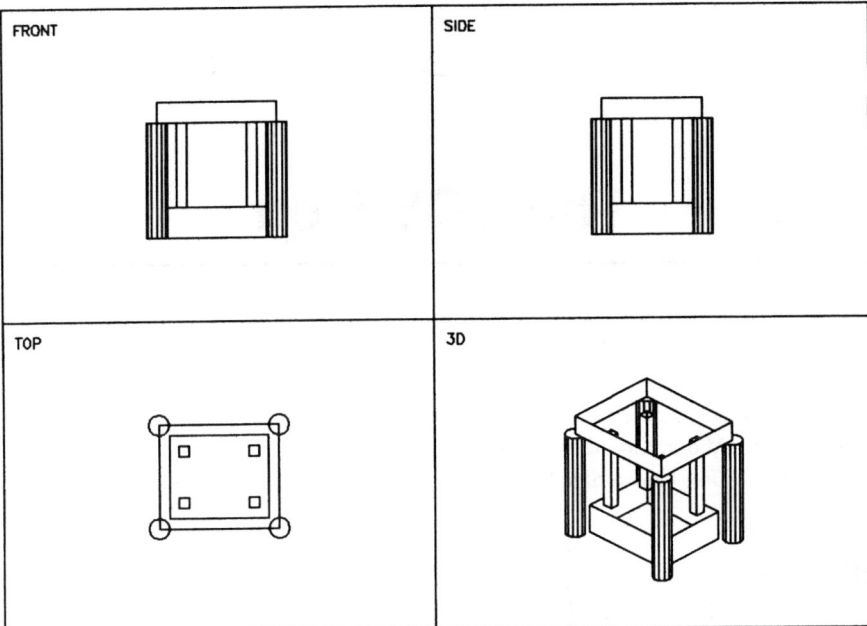

Fig. 2.6   Activity 1 – after additions.

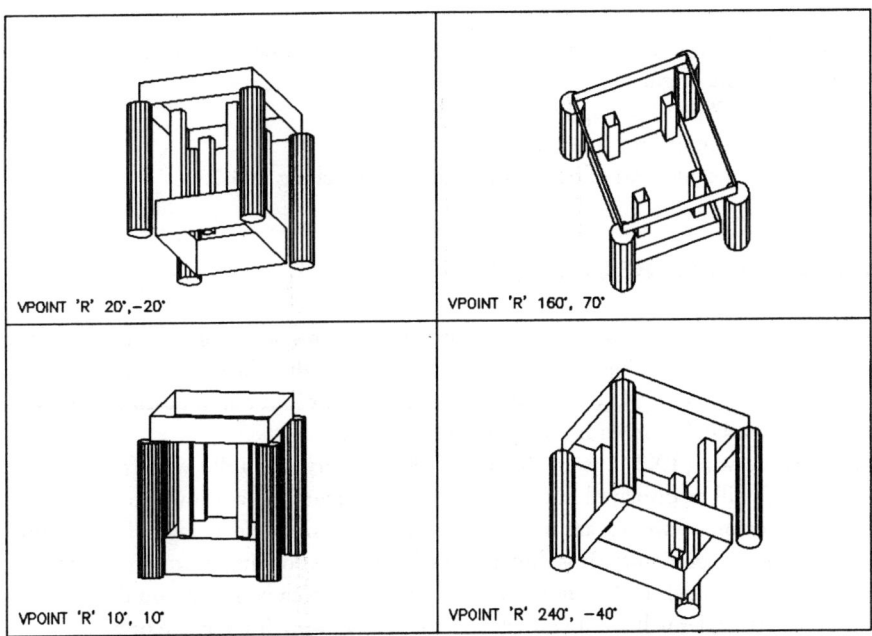

Fig. 2.7   Activity 1 – after additions with different VPOINTS (rotate option).

3

The UCS icon

There are two coordinate systems available to the user with AutoCAD:
- the WCS – World Coordinate System
- the UCS – User Coordinate System.

THE WCS

This system will be familiar to the user as it is the 'normal' system which is used with two-dimensional (2D) draughting, i.e. the origin is at the (0,0) point. The WCS is available for 3D draughting.

THE UCS

The UCS is an invaluable aid when working in 3D as it allows the user to:
- reposition the origin
- align the UCS icon to a specified 'surface'
- save UCS positions for future recall
- assist with the addition of text, dimensions and hatching in 3D.

INVESTIGATING THE UCS ICON

The UCS icon is used in 3D to assist the user with various tasks, e.g. coordinate input, text addition, dimensioning and so on, and it is essential that the user becomes familiar with its use. Without it 3D draughting would be difficult (but not impossible). We will investigate the UCS icon with a series of exercises, so:

1. Open drawing **EX3_1** from the 3DPACK directory and a 3D wire-frame model (in red) will be displayed in a four-viewport configuration as shown in Fig. 3.1.
2. Three of the viewports are in 3D and show the main disadvantage of wire-frame models, i.e. **ambiguity**. It is difficult to tell if you are looking down on, or looking up at, the model. It does not have any 'surfaces' which will help you decide. The viewport angles will tell if you whether you are viewing from above or below.

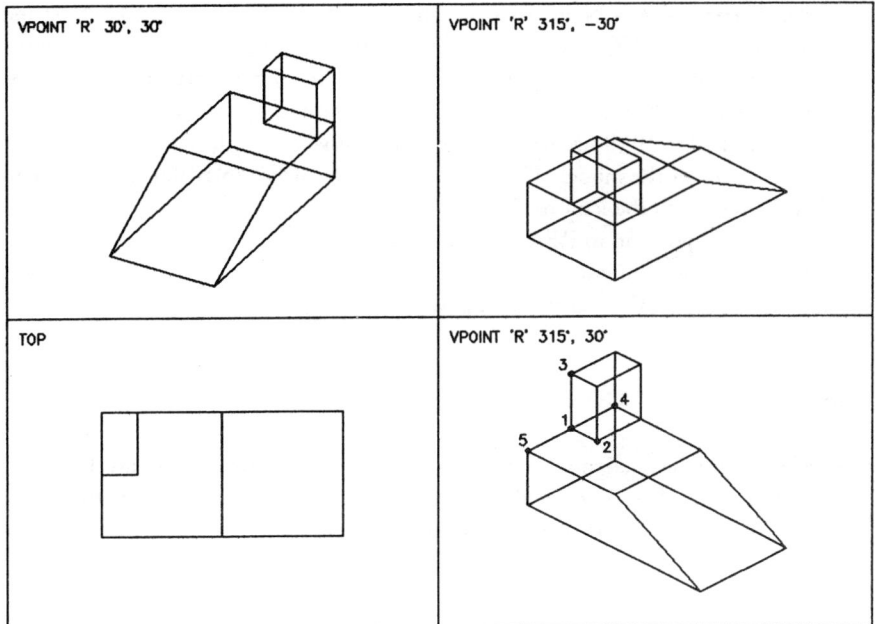

Fig. 3.1   Icon investigation.

3. From the screen menu select **SETTINGS–next**
   UCSICON–All–ON
4. The icon will be displayed in all viewports. This is the WCS icon and is orientated differently in each viewport.
5. From the screen menu select **UCSICON**
   All
   Origin
6. The icon will have 'jumped' and be aligned on the 'base' of the model in all viewports.
7. From the menu bar select **Settings**
   UCS
   Named UCS...
   prompt: UCS Control dialogue box
   respond: (a) **pick FRONT**
            (b) **pick Current**
            (c) **pick OK**
   This procedure is referred to as **restoring** a UCS position.
8. The icon will be aligned on the left vertical face of the model, and is the UCS icon. Note that:
   (a) the icon appearance in the TOP viewport. This is called the 'pencil' icon and will be displayed in a viewport when the icon is being viewed 'edge on'
   (b) the icon is the UCS icon, and differs slightly from the WCS icon. Can you spot the difference? If you cannot see any difference do not worry, it will be explained later.

9. Select **Settings–UCS–Named UCS...** and make **SLOPE** the current UCS setting the pick OK. The UCS icon will be aligned on the sloped surface in all viewports.
10. Make the bottom right viewport active (left click in it) and note the cursor cross-hairs. They are aligned with the UCS icon orientation.
11. Set the SNAP to 10 (if needed) and move the cross-hairs to the icon position at the lowest vertex of the model. The screen coordinates will display **0,0**, i.e. the origin has changed with the icon re-alignment.
12. Set the UCS position to REAR and TOP and:
    (a) note the icon orientation
    (b) check the origin is at the icon position.

THE ICON ORIENTATION

The UCS icon can be re-aligned by the user and it is normally 'attached' to one (or more) surfaces of the model being created. The appearance of the icon changes with this re-alignment, and Fig. 3.2 illustrates some of these differences.

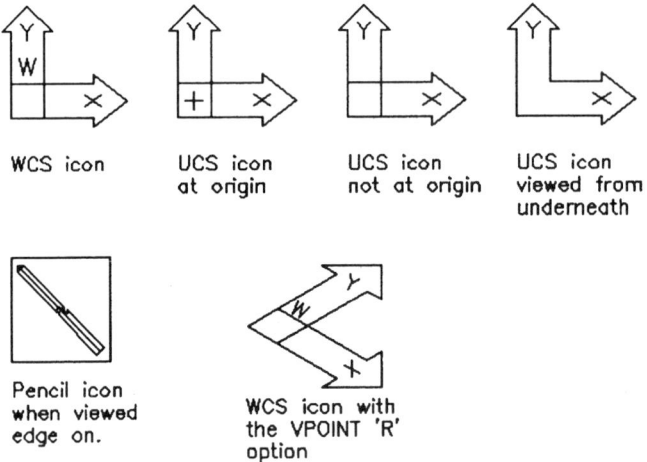

Fig. 3.2 Appearance of the WCS/UCS icon.

SAVING A NEW UCS SETTING

1. Activate the bottom right viewport – it probably is active, and toggle the SNAP off.
2. Restore UCS-TOP using the Named UCS selection.
3. From the screen menu select **UCS–next–3point**

   prompt: Origin point<0,0,0>

   respond: **pick **** INTersection then pt 1** (see Fig. 3.1)

   prompt: Point on positive portion of the X-axis

   respond: **pick **** INTersection then pt 2**

   prompt: Point on positive Y-axis portion of the UCS *XY*-plane

   respond: **pick **** INTersection then pt 3**
4. The UCS icon will:
   (a) 'jump' to point 1
   (b) be aligned with the vertical face with:
      (i) the *x*-axis pointing to point 2 and
      (ii) the *y*-axis pointing to point 3
5. From the screen menu select **UCS–Save**

   prompt: ?/Desired UCS name

   enter: **VERT1 <R>**

ROTATING THE UCS

The UCS icon can be rotated about the *X*, *Y* and *Z* axes by an amount entered by the user. The new 'rotated' position is displayed in all viewports.

1. Ensure the bottom right viewport is active with the UCS at VERT1.
2. From the screen menu select **UCS–next–X**

   prompt: Rotation about X axis

   enter: **−90 <R>**
3. The UCS icon is then aligned with:
   (a) the *x*-axis along the line 1–2
   (b) the *y*-axis along the line 1–4.
4. Select from the menu bar **Settings**

   UCS–Axis–Y

   prompt: Rotation about Y axis

   enter: **180 <R>**
5. The icon is aligned
   (a) with the *x*-axis pointing away from the model and
   (b) with the *y*-axis along line 1–4.
6. At the command line enter **UCS <R>**

   prompt: Origin/ZAxis ...

   enter: **Z <R>**

   prompt: Rotation about Z axis

   enter: **−90 <R>**
7. The icon is aligned:
   (a) with the *x*-axis pointing to point 5
   (b) with the *y*-axis pointing away from the model.
8. Each new UCS position could be saved if required.

THE UCS OPTIONS

There are several options available to the user with the UCS command. The command can be activated by three different methods, each giving the same options. The three methods are:
- from the screen menu with UCS.
- from the menu bar with Settings–UCS
- by entering UCS at the command line.

A brief description of the options is now given.

Previous	restores the previous UCS position and can be used repetitively.
Restore	the user enters the name of the required UCS position. The named UCS method can also be used, and allows the user to 'see' the names of the saved UCS positions.
Save	allows the current UCS position to be saved, the user entering the required UCS name.
Delete	removes a previously saved UCS, the user entering the name to be removed.
World	restores the WCS.
Origin	sets a new origin position, the user selecting the required origin point. The UCS will move to this new origin, and 'stay in its orientation'.
3point	sets a new origin position, the user selecting the origin point and the orientation of the X- and Y-axes.
X/Y/Z	rotates the UCS about the selected axis by an amount entered by the user.
Rename	allows a saved UCS to be renamed.
?	lists the UCS coordinates of all the saved positions, the coordinates being displayed relative to the current UCS position.

UCSFOLLOW

This is a system variable which produces an interesting effect.
1. (a) Restore UCS VERT1 with the bottom right viewport active.
   (b) Freeze layer 0 to 'remove' numbers, etc.
2. At the command line enter **UCSFOLLOW** <R>
   prompt: New value for UCSFOLLOW<0>
   enter: 1 <R>
3. Repeat step 2 in the other three viewports.
4. With the bottom right viewport active, restore UCS FRONT.
5. All viewports will display the same view of the model.
6. When UCSFOLLOW is set to 1, a 'plan' view of the model is obtained in every viewport in which UCSFOLLOW is set to 1. The word 'plan' is an AutoCAD word, and means that the view obtained will be perpendicular to the current XY-axes.
7. Restore UCS TOP and a new view will be displayed in each viewport.

This completes our investigation into the UCS icon. We will return to some of the concepts in other chapters, as well as introducing some new UCS ideas. There is no activity with this chapter.

SUMMARY

1. The UCS icon is an essential 3D draughting aid.
2. The user can reset the origin point with the UCS.
3. There are several options available with the UCS command.
4. UCS positions can be saved for future recall.
5. The UCS can be rotated about the three axes.

4

Three-dimensional coordinate input

Coordinate input in 3D can be relative to the WCS or the UCS and requires an x, y and z entry. The X-, Y- and Z-axes are mutually at right angles to each other, with the Z-axis being perpendicular to the XY-plane. If the user extends the thumb, first finger and second finger of the **right** hand and closes the other two fingers, then:
- the thumb represents the X-axis
- the first finger represents the Y-axis
- the second finger represents the Z-axis.

This representation is called the **right-hand rule** and always indicates the positions of the positive axes in relation to each other. It is always true, no matter the orientation of the UCS icon, and is very useful. Figure 4.1 demonstrates the axes orientation with two points:
- point P1 which has coordinates 50,50,20
- point P2 with coordinates −60,−30,−30.

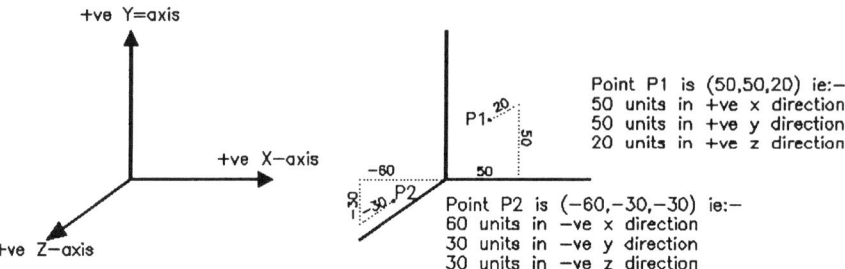

Fig. 4.1   Three-dimensional axes system with points P1 and P2.

INVESTIGATING 3D ENTITIES

1. Open drawing **EX4_1** from the 3DPACK directory which displays a four viewport layout with three red lines (Fig. 4.2):
   (a) line 1 – horizontal
   (b) line 2 – vertical
   (c) line 3 – inclined.
2. Layer OUT should be current and the bottom right viewport should be active.
3. Set the UCSICON to ON in **all** viewports – do you remember how?
4. From the screen menu select **INQUIRY**
                        LIST:
   prompt: Select objects
   respond: **pick line 1 then <R>**
5. The text screen will display the following:
   LINE   Layer: OUT
                        Space: model space
   from point, X = 100.00   Y = 0.00   Z = 0.00
   to point, X = 200.00   Y = 0.00   Z = 0.00
   Length = 100.00, angle in XY-plane = 0.0000
   Delta X = 100.00   Delta Y= 0.00   Delta Z = 0.00
6. The inquiry command gives details about the entity selected. The coordinates of line 1 are from (100,0,0) to (200,0,0), giving the length of the line as 100. The line is horizontal – angle in XY-plane is 0.
7. Repeat the INQUIRY–LIST sequence for lines 2 and 3.

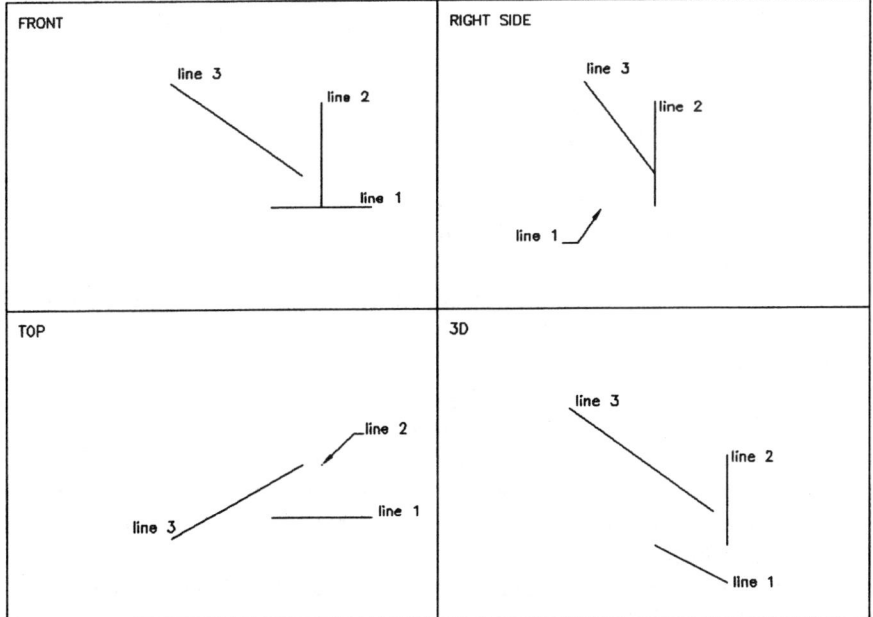

Fig. 4.2   Coordinate investigation.

ADDING ADDITIONAL LINES BY REFERENCING EXISTING ENTITIES

1. Make OBJECTS the current layer, freeze layer TEXT and thaw layer NOS1. The 3D viewport is to be active.
2. Using the LINE command:
   prompt: From point
   respond: ****ENDpoint and pick pt 1
   prompt: To point
   respond: ****ENDpoint and pick pt 4
   prompt: To point
   respond: ****ENDpoint and pick pt 2
   prompt: To point
   respond: <RETURN> to end line sequence.
3. Repeat the LINE command:
   From point: ****MIDpoint of line 1–2
   To point: ****MIDpoint of line 5–6
4. Draw a circle (CEN,RAD) with:
   centre: ****ENDpoint of pt 6
   radius: 20
5. Note the orientation of the circle in the viewports – it has been drawn on a horizontal plane.

ADDING ENTITIES BY COORDINATE INPUT

1. Restore the UCS named PT1 using the Settings–UCS–Named UCS... selection sequence.
2. Use the LINE command and at the prompts enter the following from the keyboard:
   from: 0,0,0
   to: 0,–50,0 – absolute entry
   to: @–100,0,0 – relative entry
   to: @0,0,150
   to: @100<0,0 – relative cylindrical entry
   to: @50<0<90 – relative spherical entry
   to: <R>
3. Thaw layer LET1 to display the letters a–e at the endpoints of the lines created.
4. Select from the screen menu AutoCAD–INQUIRY–ID:
   prompt: Point
   respond: ****INTersection and pick pt a
   prompt: X = 0.00   Y= -50.00   Z = 0.00
   These are the 3D coordinates of point a relative to the UCS, which is at PT1.
5. Repeat the ID command for points b–e and check the coordinate display using the entries from step 2. For example, point e: X = 0.00, Y = –50.00, Z = 200.00.
6. Restore the WCS.
7. Repeat the ID command for point a and: X = 100.00, Y = –50.00, Z = 0.00.
8. The coordinates of a point are dependent on the icon position.

ENTERING WORLD COORDINATES

Coordinates can be entered using the WCS even when the icon is positioned at a different UCS setting:
1. Restore UCS MID.
2. Using the LINE command draw:

   from: *0,0,0

   to: @*0,100,0

   to: @*0,0,100

   to: <R>
3. A horizontal and vertical line each of length 100 will be drawn from the WCS origin point. The asterisk (*) symbol 'directs' all entries to the WCS position.

UCS ENTITY

The UCS icon can be positioned relative to existing entities.
1. Make OUT the current layer.
2. Restore WCS.
3. Draw a circle (CEN,RAD), centre at 0,0,0 with radius 50.
4. Freeze layers OBJECTS, LET1 to leave the three original red lines with the numbers 1–6. The bottom right viewport should be active.
5. From the screen menu select **AutoCAD**

   UCS–next

   Entity

   prompt: Select object to align UCS

   respond: pick line 1–2 at the 1 end
6. The UCS icon jumps to point 1 and is aligned:

   (a) with the X-axis pointing towards point 2

   (b) with the Y-axis perpendicular to the line 1–2.

   Question: if icon does not 'jump'. Why?
7. At the command line enter UCS <R> then E <R> and:

   prompt: Select entity to align UCS

   respond: pick line 1–2 at the 2 end
8. Note the alignment of the UCS icon.
9. Repeat the UCS–E entry and pick any point on the circle circumference. The UCS is aligned with:

   (a) the origin at the circle centre

   (b) the X-axis pointing to the circumference pick point.
10. Repeat UCS–E and pick another point on the circle circumference, noting the alignment of the UCS.

This completes the investigation into 3D coordinate input. There is no activity, as the next chapter will investigate the creation of 3D wire-frame models, and coordinate entry will be necessary.

SUMMARY

1. Three-dimensional coordinate input can be relative to the WCS or the UCS and:
   (a) 10,20,30 – UCS entry
   (b) *10,20,30 – WCS entry.
2. Coordinate entry can be:
   (a) absolute – 10,15,20
   (b) relative – @90,80,70
   (c) cylindrical – 100<10,20
   (d) spherical – 120<20<30.
3. The UCS icon can be aligned to existing entities.

5

Creating a 3D wire-frame model

Three-dimensional wire-frame models can be created by various methods, all of which can be used in the one model construction. Some of the methods available are:
- using coordinate input – absolute or relative.
- referencing existing entities.
- using AutoCAD's editing facilities.

We will investigate some of the options available, so:
1. Open drawing file **EX5_1** from the 3DPACK directory.
2. The screen will display:
   (a) a single yellow viewport inside a black border.
   (b) six red lines on a 'black horizontal base'.
   (c) magenta numbers 1–17 at various points on the screen. These numbers are the vertices of the model to be created, and are for reference purposes only.

CONSTRUCTING THE MODEL

The wire-frame model will be constructed using the LINE command. The entry sequence which follows has been shortened, as by now I am sure that you now how to draw lines. Try and reason out the coordinate entries in relation to the lines being drawn. Remember <R> after each entry.

1. Vertical face 1–7–8–2

Line from: ****ENDpoint pt 1
      to: @0,0,70 – pt 7
      to: @200,0,0 – pt 8
      to: ****INTersection pt 2
      to: <RETURN> to end sequence.

2. Vertical face 2–8–9–4–3

Line from: ****ENDpoint pt 8
      to: @0,100,0 – pt 9
      to: ****INTersection pt 3
      to: <RETURN>
Line from: ****INTersection pt 9
      to: ****INTersection pt 4
      to: <RETURN>

3. Slope face 9–4–5–10

Line from: ****INTersection pt 9
     to: @0,60,0 – pt 10
     to: ENDpoint pt 5
     to: <RETURN>

4. Horizontal surface 7–11–12–13–10–8

Line from: ****ENDpoint pt 7
     to: @0,40,0 – pt 11
     to: @80,0,0 – pt 12
     to: @0,120,0 – pt 13
     to: ****ENDpoint pt 10
     to: <RETURN>

5. Top surface 14–15–16–17

Select from the screen menu **EDIT–COPY** and:
   prompt: Select objects
   respond: pick lines 11-12 and 12-13 then <R>
   prompt: <Base point......>/Multiple
   respond: ****INTersection and pick pt 11
   prompt: Second point of displacement
   enter: @0,0,50 <R>

Now COPY:
(a) line 14-15 from ****ENDpoint pt 15 to ****ENDpoint pt 16
(b) line 15-16 from ****ENDpoint pt 15 to ****ENDpoint pt 14.

6. Final lines

Use the LINE command (****ENDpoint?) to draw in the lines:
(a) 11–14
(b) 12–15
(c) 13–16
(d) 6–17

Your 3D wire-frame model should be complete and should resemble Fig. 5.1. The model has been constructed by referencing the entities which were available when the drawing was opened. Generally relative coordinates were used, but we also used the COPY command. At this stage save your model as **WORKDRG** in the 3DPACK directory.

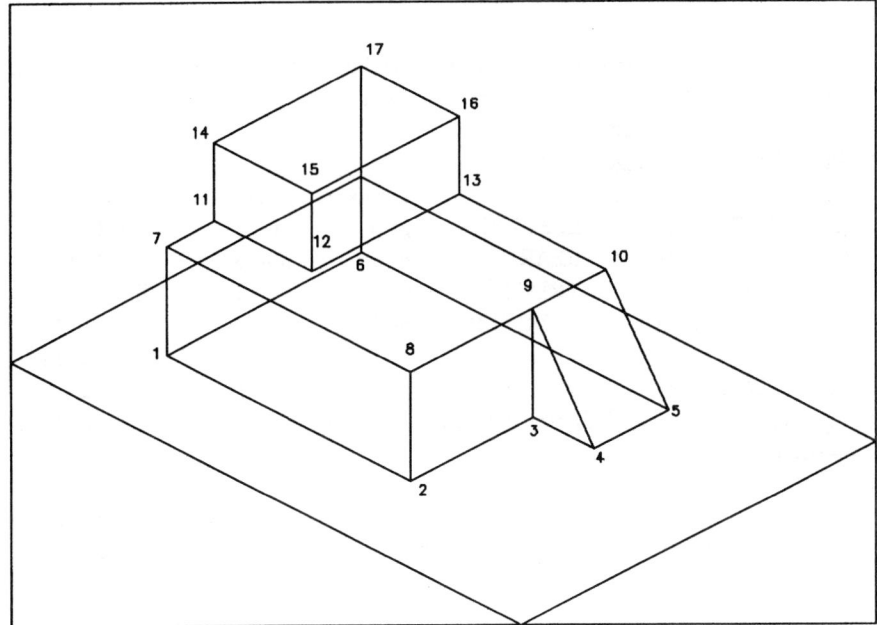

Fig. 5.1 Complete 3D wire-frame model.

UCS SETTINGS

When a 3D wire-frame model has been created, it is desirable to set and save a UCS for each surface on the model. This may be for text, dimensions, hatching, editing, etc. all of which will be investigated in later chapters.

At present the WCS is current, so:
1. Select **Settings–UCS–Named UCS...** and five UCS names will be displayed in the dialogue box, these being BASE, FRONT, RIGHT, SLOPE and TOP.
2. Make BASE current and the UCS is aligned with:
   (a) the origin at pt 1
   (b) the x-axis along line 1–2
   (c) the y-axis along line 1–6
3. Make each UCS current and check:
   (a) FRONT – aligned on surface 1–2–8–7 with origin at pt 1
   (b) RIGHT – aligned on surface 2–3–9–8 with origin at pt 2
   (c) SLOPE – aligned on surface 4–5–10–9 with origin at pt 4
   (d) TOP – aligned on surface 14–15–16–17 with origin at pt 14.

TASK

The wire-frame model has a total of 11 surfaces and five UCSs have been set and saved.
1. Create a new UCS for the six remaining surfaces.
2. Save each new UCS setting.
3. The following is a suggestion only:

Surface	Name	Origin	x-Axis	y-Axis
(a) 1–6–17–14–11–7	LEFTVERT	pt 6	line 6–1	line 6–17
(b) 6–5–10–13–16–17	REAR	pt 6	line 6–5	line 6–17
(c) 9–3–4	TRIANG	pt 3	line 3–4	line 3–9
(d) 7–11–12–13–10–8	HORIZMID	pt 7	line 7–8	line 7–11
(e) 11–12–15–14	VERT1	pt 11	line 11–12	line 11–14
(f) 12–12–16–15	VERT2	pt 12	line 12–13	line 12–15

4. I generally use the **UCS–3point** method, but it is up to you how you set the UCS positions.
5. Return to WCS.
6. Perhaps you want to save the model with the UCS settings. Remember that the name is WORKDRG and you want to update the existing drawing of that name.

ADDING OBJECTS TO THE MODEL

The wire-frame model consists of lines drawn on layer OUT in red. We now want to add some other features to the model – holes and slots – and will add these on a different layer to differentiate them from the outlines.

1. Circles on the top and base

1. Ensure WCS is current.
2. Make layer OBJECTS (blue) current.
3. Draw a circle (CEN,RAD) with:
    (a) centre at point 40,70
    (b) radius 25.
4. Restore UCS BASE, and repeat the circle command entering the same centre point and radius.
5. Restore UCS TOP and again repeat the circle command – same values.
6. Three blue circles have been created, each in a different position which is dependent on the icon position:
    (a) WCS – circle is on the 'black border' at 40,70 from the WCS origin at (0,0)
    (b) UCS BASE – circle centre is at 40,70 from pt 1
    (c) UCS TOP – circle centre is at 40,70 from pt 14.
7. The desired circle is the one created with UCS TOP, so ERASE the first two circles drawn.
8. We require another blue circle on the base, directly under the one just drawn. This could be achieved by setting UCS BASE and then drawing the circle using coordinate input. We will however use the COPY command to obtain this additional circle. (Aside: what would be the base circle centre coordinates if the UCS BASE was restored?)

9. UCS still at TOP?
10. Select **EDIT–COPY** and:
    prompt: Select objects
    respond: **pick blue circle then <R>**
    prompt: <Base point...
    respond: ******CENtre and pick the circle**
    prompt: Second point of displacement
    enter: **@0,0,–120 <R>** – why these coordinates?

2. Square on horizontal surface

1. Restore UCS HORIZMID (origin at pt 7).
2. From the menu bar select **Draw**
                         Rectangle
    prompt: First corner
    enter: **100,125 <R>**
    prompt: Other corner
    enter: **140,85 <R>**
3. A 40-unit blue square will be drawn alongside line 12–13.
4. COPY the square: (a) from one vertex and
                      (b) by @0,0,–15
5. Draw in the four vertical lines required to give the 'square cut-out' effect.

3. Triangle shape on slope

1. Restore UCS SLOPE
2. With LINE, draw
    from: 15,15
        to: @30,0
        to: @ 0,55
        to: close.
3. Return to WCS.
4. At this stage the additional items have been added to the model and it should be saved as WORKDRG for future use. Your screen display should resemble Fig. 5.2.

SUMMARY

1. Wire-frame models can be constructed by:
    (a) coordinate input
    (b) referencing existing entities
    (c) using AutoCAD's editing commands.
2. The UCS is **important** in positioning entities.
3. UCSs should be set and saved for all surfaces on a wire-frame model.

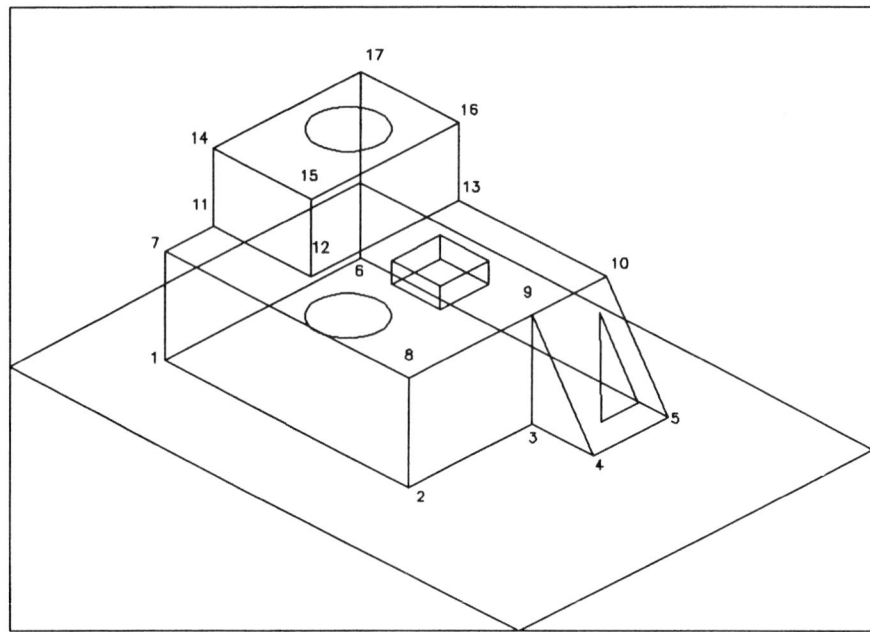

Fig. 5.2 WORKDRG with objects added.

ACTIVITY

Open drawing **ACT_2** from the 3DPACK directory to display Fig. 5.3 with:
(a) a three-viewport configuration of 3D, TOP and FRONT views
(b) a red hexagon in each view
(c) numbers 1–6 relating to the hex vertices
(d) the WCS at the hex 'centre'
(e) layer OUT current and the 3D viewport active.

1. Using the following vertical line lengths, draw an irregular truncated hexagonal prism with layer OUT current:

length at vertex 1:	50	length at vertex 4:	150
length at vertex 2:	50	length at vertex 5:	150
length at vertex 3:	80	length at vertex 6:	80

(Hint: line from ****INT pt 1 to @0,0,50, etc.)

2. With OBJECTS layer current (blue) add the seven(?) lines making up the top surfaces of the prism, to give Fig. 5.4.

3. Set and save six new UCS positions, one for each vertical face (for example):

Face	Origin	x-Axis	y-axis	Name
1–2	pt 1	along line 1–2	vertical	FACE1
2–3	pt 2	along line 2–3	vertical	FACE2
3–4	pt 3	along line 3–4	vertical	FACE3
etc.				

4. Restore WCS.

5. Save model as **ACTDRG** in 3DPACK directory for future activities.

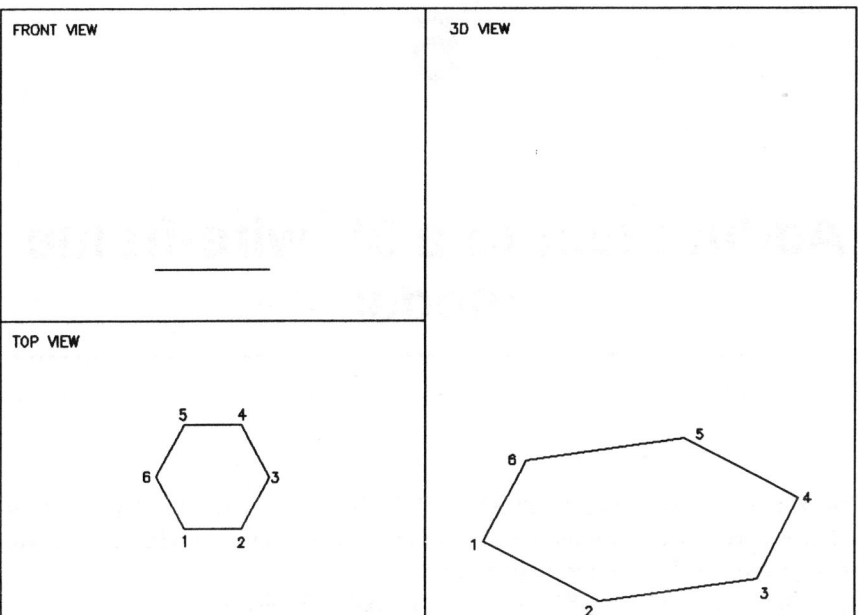

Fig. 5.3 Activity 2 – 3D coordinate activity (original).

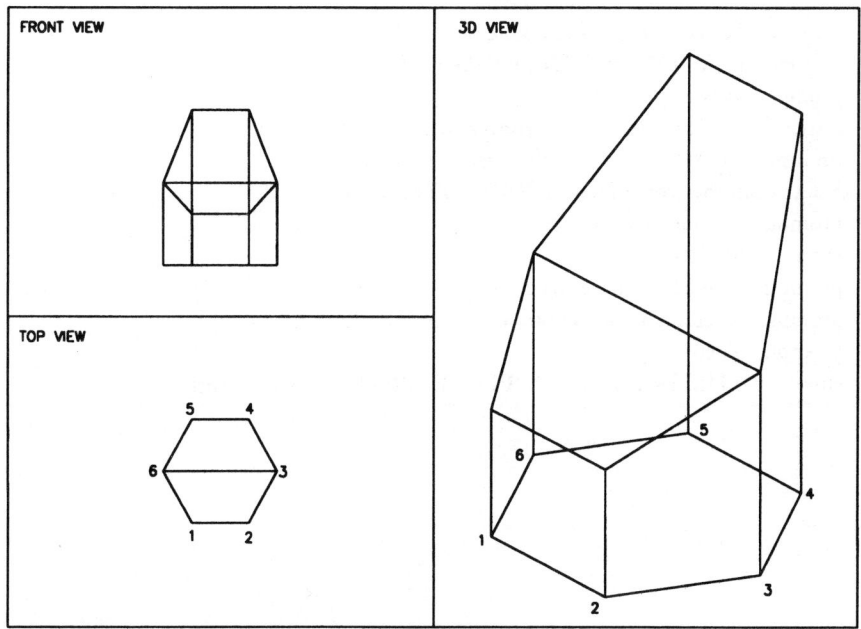

Fig. 5.4 Activity 2 – 3D coordinate activity (completed).

6

Adding text to a 3D wire-frame model

Our existing wire-frame model (called WORKDRG) has been created with red outlines and blue objects. We now want to add text to the 'surfaces' of the model, and this will require setting the UCS to the desired surface.

1. Open the 3D model WORKDRG from the 3DPACK directory.
2. Check
   (a) that WCS is current
   (b) that the layer is current? – mine was OBJECTS.
3. Make layer TEXT (green) current.
4. Select from the menu bar **Draw**

   Text

   Set Style...

   prompt: Select Text Font dialogue box

   respond: **pick ROMAN DUPLEX then OK**

   prompt: various prompts ...

   respond: **<RETURN> to all prompts** i.e. accept the defaults.

   prompt: ROMAND is now the current text style.
5. Select from the screen menu **DRAW–DTEXT** and:

   prompt: ...<Start point>

   enter: **50,25 <R>**

   prompt: Height<?> and enter **8 <R>**

   prompt: Rotation angle<0.0000> and enter **0 <R>**

   prompt: Text

   enter: **THIS IS THE BASE 'PLANE'** <R><R> – two returns!

TEXT ON THE MODEL 'SURFACES'

With AutoCAD, text can only be added to the *XY*-plane, which is dependent on the UCS position and orientation. This requires that the correct UCS is restored to the surface on which the text is to be added.

1. Restore UCS FRONT (Settings–UCS–Named UCS...).
2. Using the DTEXT command, enter the following:

start point:	30,10
height:	10
rotation:	0
text:	FRONT

3. The word 'FRONT' will be displayed with Roman Duplex text font, at a height of 10 on the left vertical face.

TASK

Text has to be added to **all** surfaces of the 3D model. As the addition of text is a fairly easy process, I've decided that a different text font should be used for each surface.

The following table gives the required text font, height, rotation angle and actual text which is to be added to the surfaces of the model. The procedure is the same for each text item:

1. Restore the desired UCS.
2. Set the required text font.
3. Use the DTEXT command to add the text.

    The table of text fonts is:

UCS	Start point	Text font	Height	Rotation	Actual text
BASE	100,20	Italic Triplex	15	0	BASE
HORIZMID	10,10	Script Complex	10	0	MID
LEFTVERT	100,5	Monotxt	8	15	LEFT
REAR	30,50	Sansserif	20	0	REAR
RIGHT	10,10	Gothic English	10	5	RIGHT
SLOPE	5,5	Technic	10	0	SLOPE
TOP	10,10	City Blueprint	25	5	TOP
TRIANG	5,5	Txt	10	0	TRI
VERT1	10,10	Romantic Bold	30	0	V1
VERT2	10,10	Superfrench	30	0	V2

When complete, your model should resemble Fig. 6.1.

Restore the WCS, and save your model as **WORKDRG**.

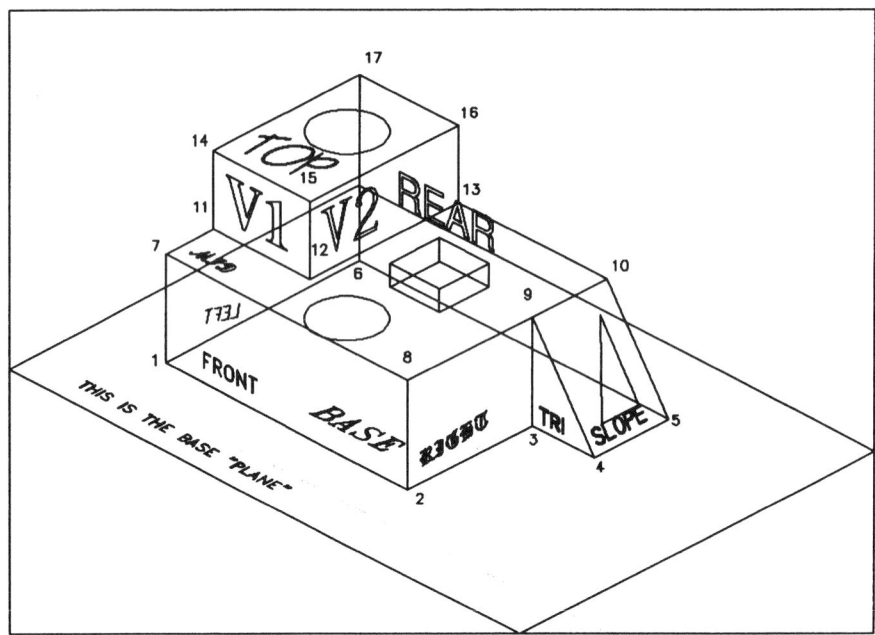

Fig. 6.1   WORKDRG with TEXT added to surfaces.

SUMMARY

1. Text which is to be added to any surface of a 3D model, must have the appropriate UCS set.
2. Different text fonts can be used if required.

ACTIVITY

Recall the activity drawing **ACTDRG**, refer to Fig. 6.2, and add eight items of text to the surfaces.

Note

1. Select your own text font.
2. The same text font is to be used for all text items.
3. Pick suitable start points (centre option?), heights and rotation angles.
4. Two new UCSs will be needed for the sloped surfaces, but you should be able to manage this(!?)
5. When complete, save your drawing as **ACTDRG**.

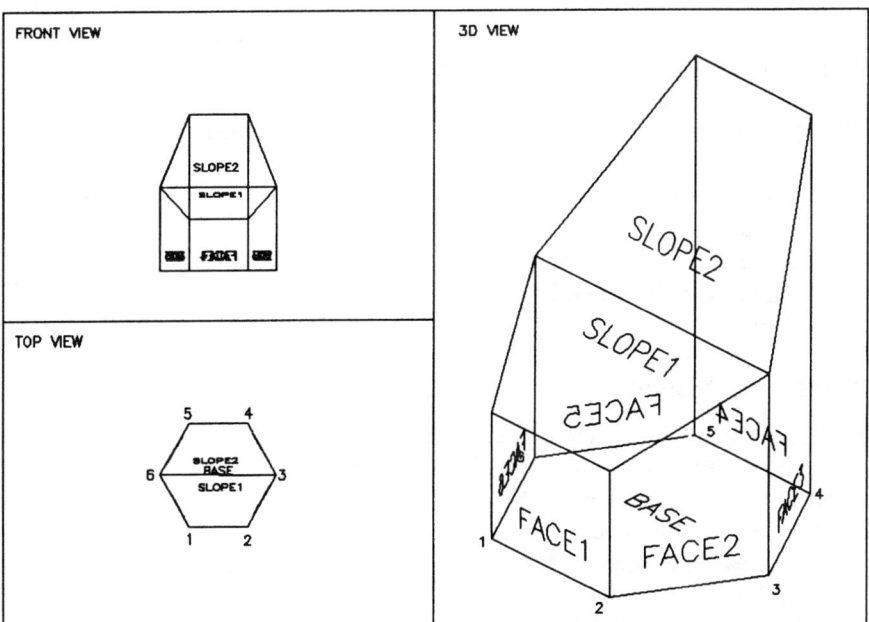

Fig. 6.2 ACTDRG with text added.

7

Dimensioning a 3D wire-frame model

Dimensions can be added to a 3D wire-frame model, but as with text, the position of the UCS icon is important. To demonstrate how dimensions are added:

1. Open drawing **WORKDRG** from the 3DPACK directory.
2. Make DIMENS the current layer, and freeze layer TEXT. Also freeze the 'base plane' layer 0.
3. Alter the following DIMVAR values (if necessary):
   - (a) dimasz: 6
   - (b) dimcen: 0
   - (c) dimtad and dimtih: OFF
   - (d) dimtxt: 8

1. WCS dimensions

1. Restore the WCS – it should be current?
2. Using the DIM command:
   - (a) horizontal dimension line 1–2
   - (b) vertical dimension line 1–7
   - (c) vertical dimension line 2–4
   - (d) radius dimension the top circle.
   Note: use OSNAP END or INT to assist with selection.
3. The result of these dimensions is Fig. 7.1(a).
4. Study the added dimensions then erase them.

2. UCS BASE dimensions

1. Restore UCS BASE.
2. Dimension the same four entities as before, i.e.
   - (a) horizontally line 1–2
   - (b) vertically line 1–7
   - (c) vertically line 2–4
   - (d) radially the top circle.
3. Study the results – Fig. 7.1(b), then erase the four dimensions.

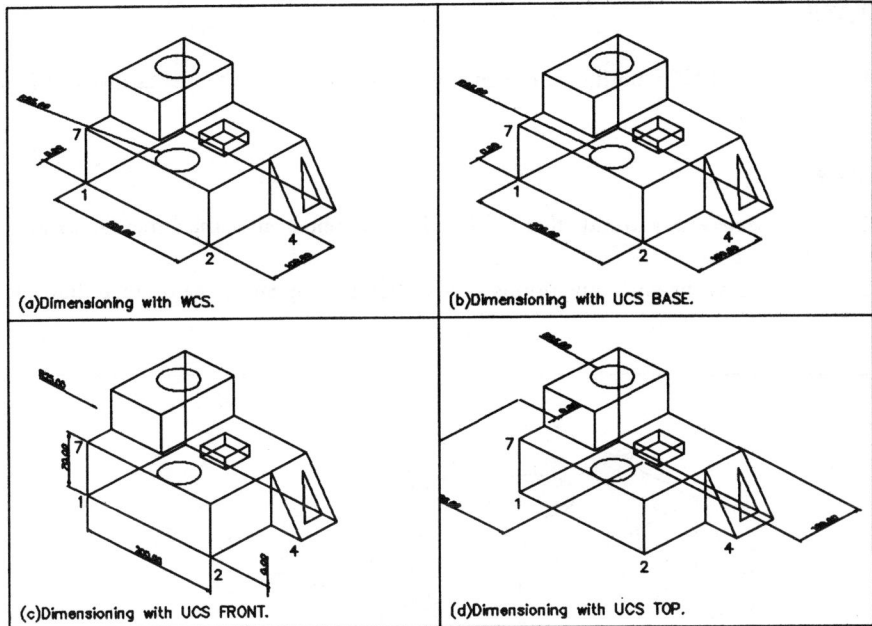

Fig. 7.1 WORKDRG with dimensions added using different UCS positions.

3. UCS FRONT dimensions

1. Restore the UCS FRONT position.
2. Dimension the same four entities as before – Fig. 7.1(c).
3. Erase the dimensions.

4. UCS TOP dimensions

1. Restore UCS TOP.
2. Dimension the four entities to give Fig. 7.1(d).
3. Erase the dimensions.

ANALYSIS OF THE DIMENSIONING

The completed exercise has proved that it is possible to dimension a 3D wire-frame model, but the actual added dimensions are orientated dependent on the UCS position. The correct dimensioning was obtained with the following:
   (a) horizontal line 1–2: with WCS, UCS BASE and UCS FRONT
   (b) vertical line 1–7: with UCS FRONT
   (c) vertical line 2–4: with WCS and UCS BASE
   (d) top circle: with UCS TOP.

TASK

Using the saved UCS positions, dimension the 3D wire-frame model as given in Fig. 7.2.

Note

1. It may be necessary to add other UCS positions to obtain all of the dimensions shown in Fig. 7.2.
2. Save the drawing with dimensions as **WORKDRG**, updating the original drawing.

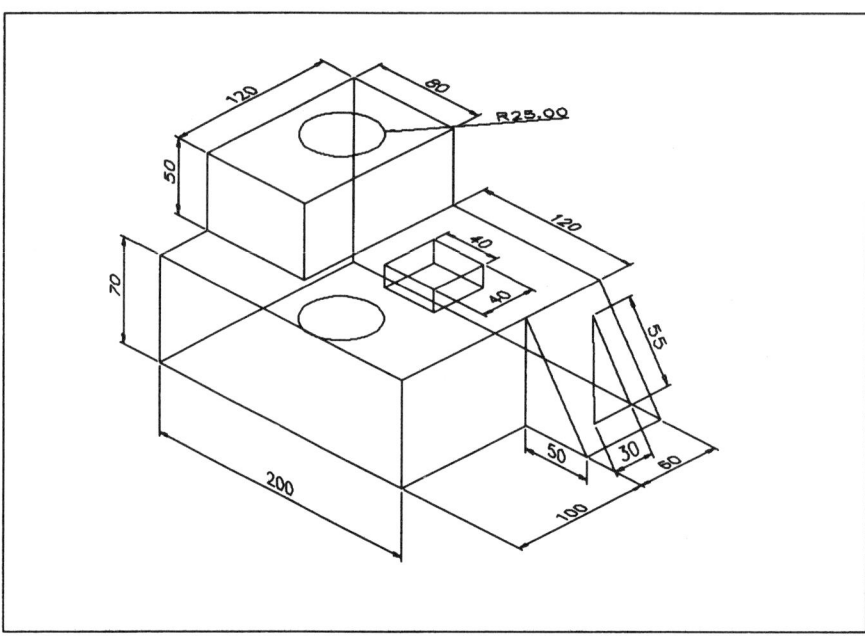

Fig. 7.2   WORKDRG with required dimensions added.

SUMMARY

1. Dimensions can be added to 3D wire-frame models.
2. The dimension orientation is dependent on the UCS position.
3. Horizontal dimensions are **always** in the icon *x*-axis direction.
4. Vertical dimensions are **always** in the icon *y*-axis direction.
5. Circles can only be dimensioned with the UCS correctly set.
6. With linear dimensioning, the baseline and continue options are available.
7. Leader dimensioning is permissible, but the result may not be as expected.
8. It may be necessary to alter some of the **DIMVAR** values to obtain a 'clearer' dimension effect.
9. The use of OSNAP (END or INT) is recommended when dimensioning.

ACTIVITY

1. Open drawing **ACT_3**.
2. Refer to Fig. 7.3 and add the given dimensions. *Note*: UCS changes will probably be required.
3. Save as required.

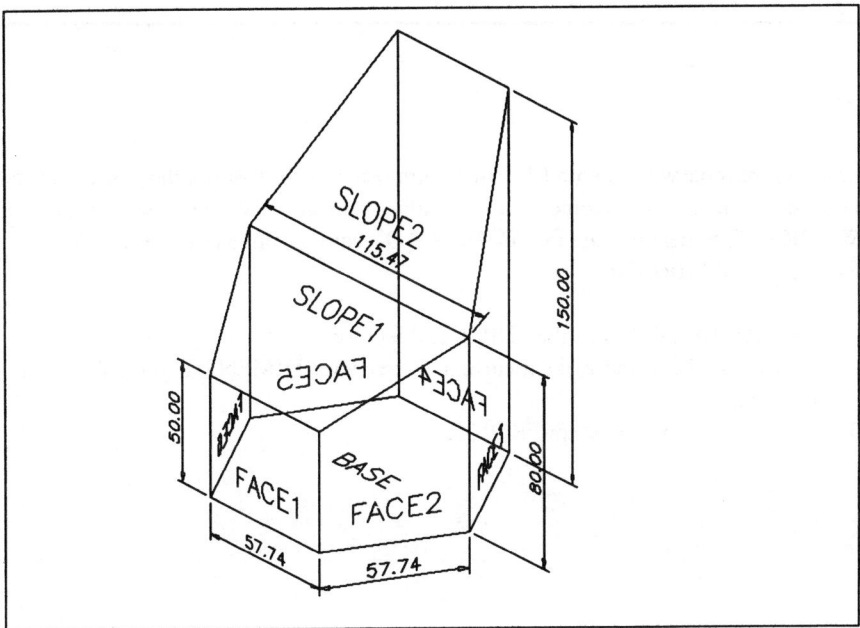

Fig. 7.3   Activity 3 – adding dimensions.

8

Editing in 3D

Editing is possible with 3D models, but as with text and dimensions, the position of the UCS determines the outcome of the editing command. We will use our existing WORKDRG to demonstrate four of AutoCAD's editing commands: COPY, ARRAY, ROTATE and MIRROR.

1. Open **WORKDRG** from the 3DPACK directory.
2. Thaw layer TEXT and make it current. Freeze layer DIMENS. Layer 0 should still be frozen?
3. Erase all items of text except FRONT.

COPY COMMAND

1. Restore UCS BASE.
2. From the screen menu select **EDIT–COPY** and:
   prompt: Select objects
   respond: **pick the green text item (FRONT) and the four red lines**
   1–2, 2–8, 8–7, 7–1 then <R>
   prompt: <Base point ...
   respond: **pick INT pt 1**
   prompt: Second point of displacement
   enter: **@0,–100**
3. Restore the named UCS FRONT.
4. COPY the same five entities (1 text and 4 lines) as step 2 and:
   (a) pick INT pt 1 as the base point
   (b) enter @0,–100 as the second point of displacement.
5. The result of the identical COPY effect with two different UCS settings is shown in Fig. 8.1.
6. You should be able to reason out the two COPY effects from the coordinate input. Remember that these are relevant to the current UCS position.
7. Erase the two copy effects.

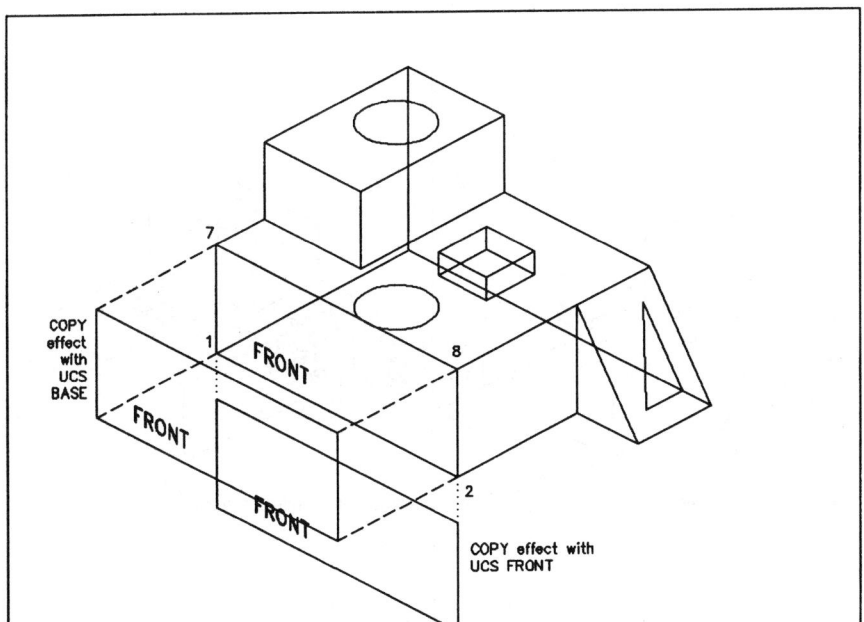

Fig. 8.1 COPY effect on 3D wire-frame model.

ARRAY

1. UCS FRONT current.
2. From the screen menu select **EDIT–ARRAY** and at the various prompts enter the following:

Select objects:	pick the FRONT text item then <R>
Rect/Polar:	P
Center point:	pick INT pt 1
Number of items:	12
Angle:	360
Rotate:	Y

3. Study the resultant array and how it has been created then enter U <R> to undo the array command.
4. Restore UCS BASE and repeat step 2 as given.
5. The two array effects are shown in Fig. 8.2. Again you should be able to reason out these arrays from the current UCS position. Remember that the ARRAY command is created on the *XY*-plane orientation.
6. Erase (or U) the array.

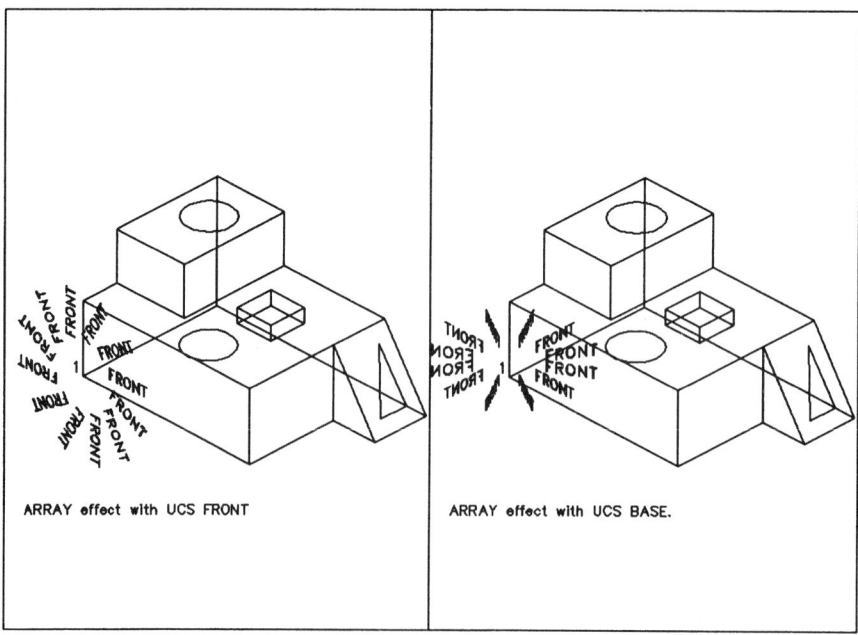

ARRAY effect with UCS FRONT   ARRAY effect with UCS BASE.

Fig. 8.2   ARRAY effect on 3D wire-frame model.

ROTATE

1. UCS FRONT current.
2. From the screen menu select **EDIT–next–ROTATE** and:
   prompt: Select objects
   respond: **window the 12 blue lines which make up the 'square hole' then <R>**
   prompt: Base point
   enter: **0,0 <R>**
   prompt: Rotation angle
   enter: **−90 <R>**
3. Study the rotation then enter **U <R>** to undo it.
4. Restore UCS BASE and repeat the ROTATE command as step 2:
   (a) picking the same 12 lines
   (b) entering 0,0 and −90 as the base point and angle, respectively.
5. The effect of the two rotations is shown in Fig. 8.3.
6. Erase the rotate effect.

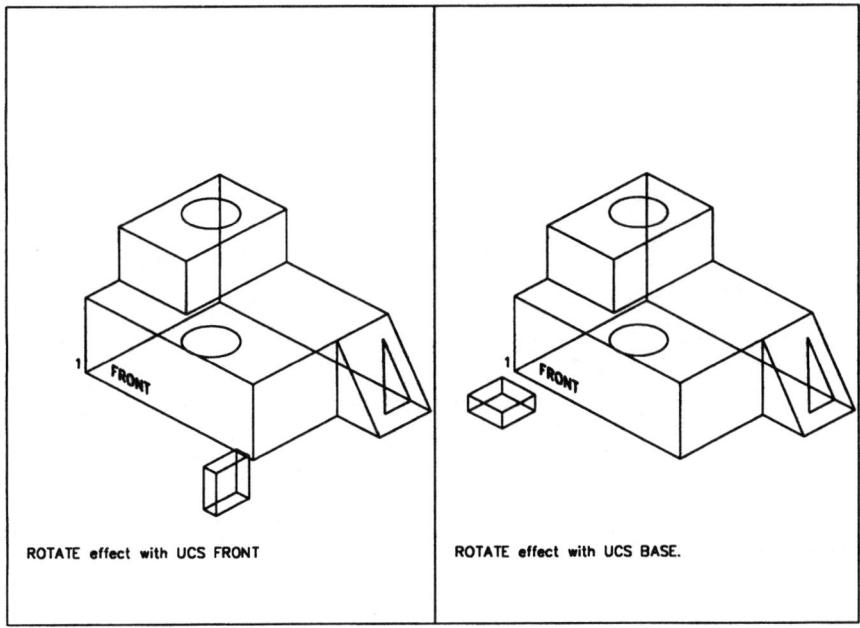

ROTATE effect with UCS FRONT

ROTATE effect with UCS BASE.

Fig. 8.3   ROTATE effect on 3D wire-frame model.

MIRROR

1. UCS BASE current.
2. With the MIRROR command:
   prompt: Select objects
   respond: **pick the 3 blue lines of the triangle then <R>**
   prompt: First point on mirror line
   respond: **pick INT pt 1**
   prompt: Second point
   respond: **pick INT pt 2**
   prompt: Delete old objects
   enter: **N <R>**
3. Observe the mirror effect then **U <R>**
4. Restore UCS FRONT and repeat the MIRROR command of step 2:
   (a) picking the three blue triangle lines
   (b) picking the same two mirror line points
   (c) entering N to the delete objects prompt.
5. The effect of the two mirror commands is shown in Fig. 8.4.
6. Do not save any of these edited drawings, i.e. leave WORKDRG as it was when it was opened at the start of the exercise.

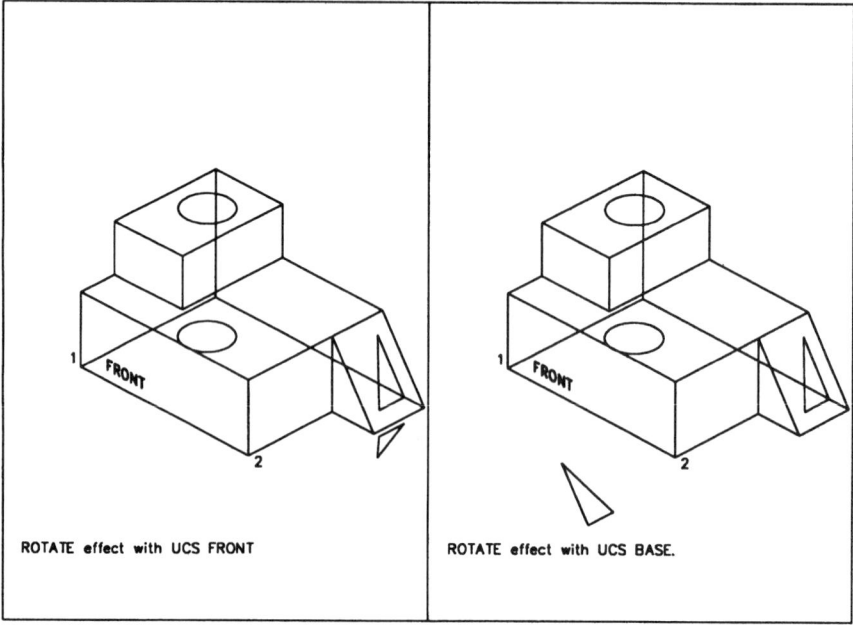

Fig. 8.4   MIRROR effect on 3D wire-frame model.

Note

The four commands demonstrated are all 2D editing commands. All 2D editing commands are available for use with 3D models, but the user should be aware that certain 'messages' will be displayed in the command prompt area with several of these commands:

- **1 was not parallel to the UCS**

  This message means exactly what it says. Entities have been picked for editing which are not on (or parallel) to the current UCS position. The editing effect will not work with the selected entity.

- **View is not plan to UCS. Command results may not be obvious**

  You are about to use an edit command, and the result may not be as expected. This message is displayed with BREAK, EXTEND, OFFSET, TRIM. You can still attempt to edit.

SUMMARY

1. The 2D editing commands can be used with 3D models.
2. The position of the UCS is important.
3. Some editing results may not be as expected, due to the position of the UCS.
4. There are three commands specific to 3D models, these being: ARRAY 3D, MIRROR 3D, ROTATE 3D. These commands will be discussed in a later chapter.

ACTIVITY

No activity for this section.

9

Hatching in 3D

Both the HATCH and the BHATCH commands are available with 3D wire-frame models, but (as usual) the UCS position is critical. The 'surface' to be hatched must have the 'correct UCS setting'.

It is possible to:

- hatch directly onto 'surfaces'
- add hatching to 'faces'
- hatch with 3D Polylines.

At present we will only consider the first option – adding hatching directly to the required surfaces. Faces and 3D polylines will be considered in later chapters.

1. Open **WORKDRG** from the 3DPACK directory – with dimensions?
2. Use Layer Control to:
   (a) make SECT the current layer
   (b) freeze layer DIMENS
   (c) thaw layer 0 – the base plane.

WCS HATCHING

Suppose we want to add hatching to the TOP surface, then:

1. WCS current.
2. Select **DRAW–HATCH** from the screen menu and enter:

   Pattern: U

   Angle: 45

   Spacing: 5

   Double: N

   Select objects: pick the four red lines on the top surface then <R>
3. Hatching is added to the drawing, but not where expected. It is on the 'base plane' and this is due to the WCS.
4. Erase the hatching.

UCS TOP HATCHING

1. Restore UCS TOP.
2. Repeat the HATCH command, accepting the U, 45, 5 and N defaults, then:
   prompt: Select objects
   respond: **pick the four red lines and blue circle on top surface then <R>**
3. Hatching will be added as required.

UCS SLOPE HATCHING

1. Restore UCS SLOPE.
2. Use the HATCH command and:
   (a) accept all defaults
   (b) pick the four red lines and three blue triangle lines on the slope surface then <R>
3. Hatching as expected?

UCS FRONT HATCHING

1. Restore UCS FRONT.
2. From the menu bar select **Draw**
                    Hatch...
   prompt: Boundary Hatch dialogue box
   respond: (a) pick Hatch Options...
            (b) pick Stored Hatch Pattern 'radio button'
            (c) pick Pattern...
            (d) pick Next, Next, Next
            (e) pick **stars** hatch pattern icon
            (f) alter Scale to 2 and Angle to 0
            (g) pick OK
            (h) pick **'Pick Points <'**
            (i) pick any point within the front vertical face then <R>
            (j) pick Preview
            (k) observe hatching then <R>
            (l) pick Apply
3. A star hatch pattern is added to the front vertical surface.

TASK

Use the BHATCH command with the patterns, scale and angles given below to add hatching as shown in Fig. 9.1.

UCS	Hatch pattern	Scale	Angle
HORIZMID	HEX	2	0
RIGHT	ANGLE	2	45
TRIANG	DOTS	2	0
VERT1	SQUARE	2	45
VERT2	SQUARE	2	−45

When complete, restore WCS and save model as WORKDRG in the 3DPACK directory.

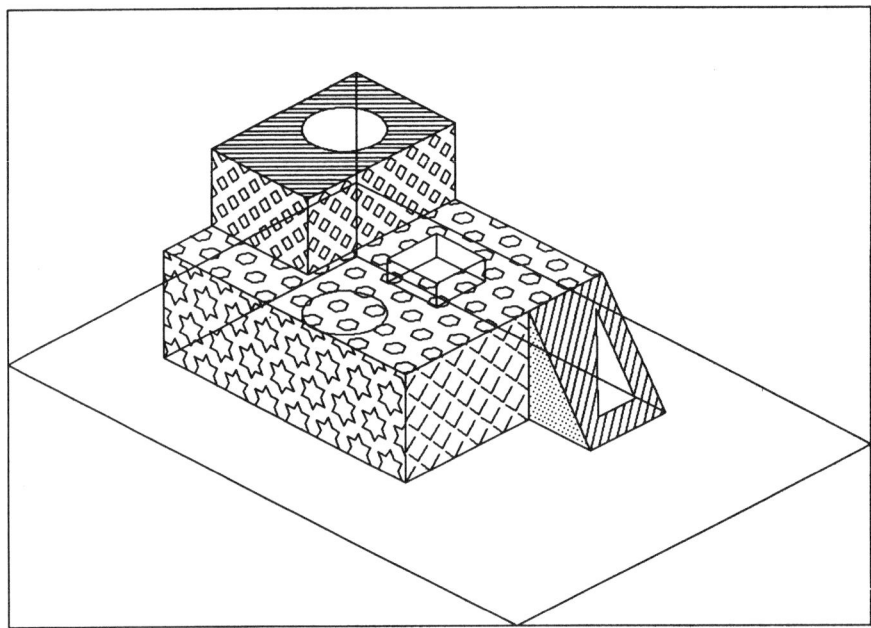

Fig. 9.1   WORKDRG with HATCHING added to several 'surfaces'.

SUMMARY

1. Hatching can be added to the 'surfaces' of a 3D wire-frame model.
2. The UCS position is critical when adding hatching.
3. Both HATCH and BHATCH commands are available.
4. Remember that hatching can use 'a lot of memory'.

ACTIVITY

Open the **ACT_4** drawing from the 3DPACK directory, and a regular hexagonal pyramid will be displayed in a three-viewport configuration.

Using the three saved UCS positions, add hatching to the three sloped surfaces. The hatching can be your own, but Fig. 9.2 gives my hatch pattern which was TRIANG, with a scale of 1 and an angle of 0.

As this activity is displayed in multiple viewports, you will 'see' the hatching being added to the sloped surfaces in the other two non-active viewports.

Save if required, but we will not refer to the diagram.

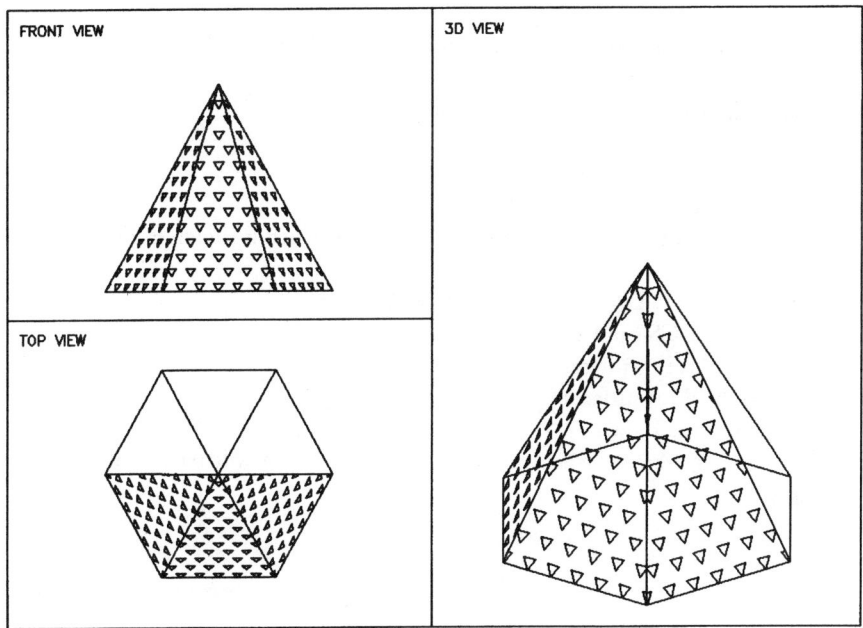

Fig. 9.2   Activity 4 with hatching added to three surfaces of the pyramid.

10

User exercise 1

We have now investigated different aspects of 3D wire-frame model creation:
  (a) constructing using coordinate input
  (b) setting, saving and restoring UCS positions
  (c) adding text to surfaces
  (d) dimensioning the model
  (e) adding hatching.
As a check on your 'progress', I have set you a task which is shown in Fig. 10.1, so:

1. Open drawing **USEX1** from the 3DPACK directory.
2. Using the set UCS's provided:
   (a) add the three text items
   (b) hatch the three surfaces
   (c) dimension as shown.
3. Save you user exercise, when complete, as USEX1 – we will refer to it in a later chapter.

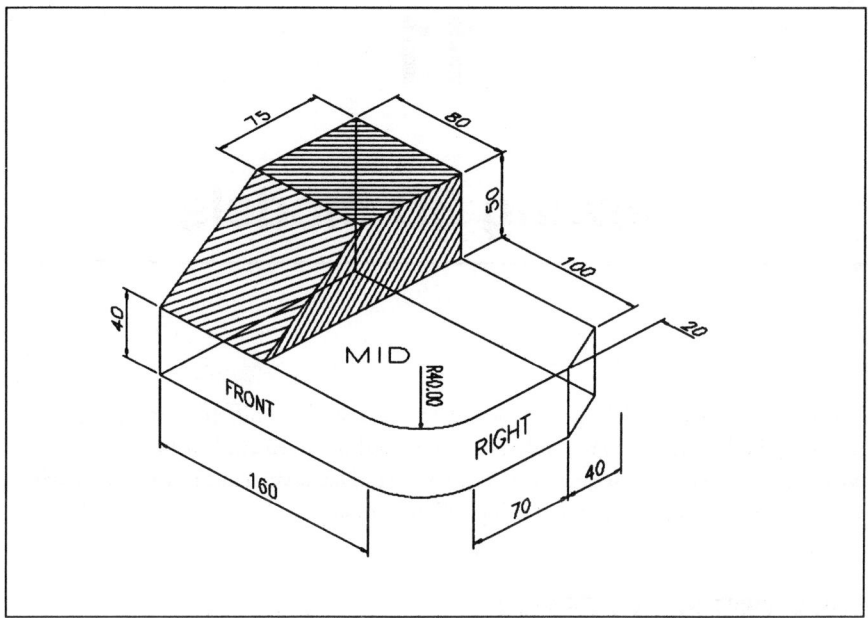

Fig. 10.1   User exercise 1.

11

Viewing 3D models

Three-dimensional models can be viewed by the user from different 'standing points' by using the **VPOINT** command. This command has been used in previous chapters without any attempt to investigate it. The command has several options available, which will now be investigated in detail using different examples.

THE VPOINT ROTATE OPTION

1. Open drawing **EX11_1** from the 3DPACK directory.
2. The screen will display several red lines in 'plan' view?
3. From the screen menu select **DISPLAY**

   VPOINT:

   rotate

   prompt: Enter angle in *XY*-plane from *X*-axis<?>
   enter: **315** <R>
   prompt: Enter angle from *XY*-plane<?>
   enter: **30** <R>
4. The wire-frame model will be displayed in a 3D orientation, and is a cuboid shape with two 'ramps'.
5. *Question*:   are you looking down or up at the model?
   Answer: (1) you cannot really tell from the model
                       (2) icon orientation tells that you are looking from above
                       (3) how do we know this?
6. Repeat the **VPOINT–rotate** selection and enter:
   (a) 315 as the angle in *XY*-plane – first prompt
   (b) −30 as the angle form *XY*-plane – second prompt.
7. The model display changes as does the icon. We are now looking at the model from below – no 'box' in icon.
8. At the command line enter **VPOINT** <R>
   prompt: Rotate/<View point><.......>
   enter: **R** <R> – for rotate option
   prompt: Enter angle in *XY*-plane ... and enter **30** <R>
   prompt: Enter angle from *XY*-plane ... and enter **60** <R>
9. The resultant display is the model viewed from above at a 'steeper' angle.

10. Now using   (a) DISPLAY–VPOINT–rotate from the screen menu

        or   (b) VPOINT-R from the keyboard

enter the following angle values in response to the first and second prompts:

- angle in *XY*-plane from *X*-axis   0   60   120   80   300
- angle from *XY*-plane   10   40   90   −30   −300

11. Observe the icon orientation after each entry.
12. Refer to Fig. 11.1 for the expected display from all the entered VPOINT angles.
13. Do not save your changes.

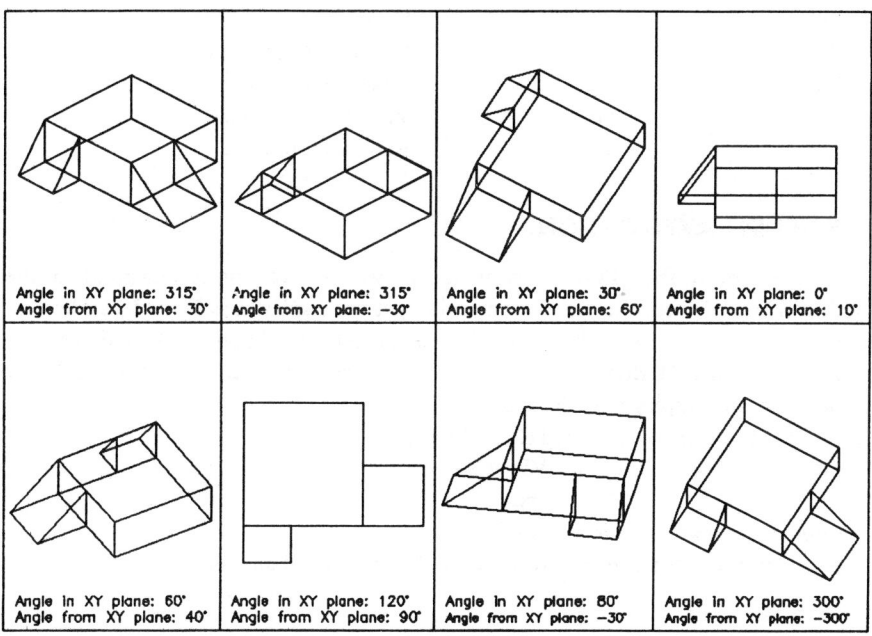

Fig. 11.1 Various VPOINT 'R' entries with EX11_1.

THE VPOINT ROTATE ANGLES

The rotate option of the VPOINT command requires the user to enter two angle values from the keyboard, these being:

1. angle in *XY*-plane from X axis : this is the user's 'stand point' on the horizontal *XY*-plane looking towards the model. Imagine you are walking around the model from 0° at the right-hand side, through a complete 360° circle. The model appearance will change as you move, and:

   0 – view from right side

  90 – view from the 'back' of model

 180 – view from left side

 270 – view from model 'front'

 360 – same as 0

2. angle from *XY*-plane : this is the angle at which the user looks at the model and can be positive (+) – from above; or negative (–) – from below. The greater the entered angle, the steeper is the user's 'angle of inclination' and:

   0 – viewed from horizontal

   90 – viewed vertically down

   –90 – viewed vertically upwards.

3. Some useful entries for the angles are:

First angle	Second angle	View
0	0	View from model right side (end view)
90	0	View from rear
180	0	End view from left side
270	0	Front view of model
0	90	View from top (plan)
0	–90	View from bottom.

THE VPOINT AXES OPTION

The axes option of the VPOINT command allows the user infinite viewpoints of the model. When selected, the axes tripod and target are displayed as Fig. 11.2. The user moves the cross-cursor into the required circle quadrant to obtain the desired view.

1. Open the USEX1 drawing (created in the previous section) from the 3DPACK directory and display only the model.

2. From the screen menu select **DISPLAY**

   VPOINT:

   axes

   prompt: tripod/target icons as Fig. 11.2

   respond: move the cursor to quadrant A (see Fig. 11.3) and left click.

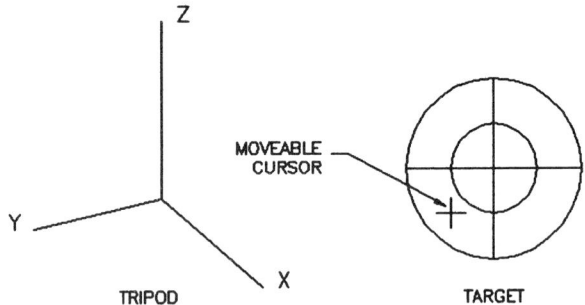

Fig. 11.2   The VPOINT tripod and target.

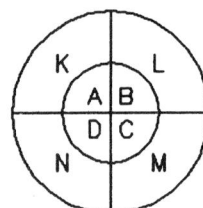

Fig. 11.3   Selection points.

3. The model will be displayed at this selected setting. Observe the icon – looking from above.
4. Select from the menu bar **View**
   > **Set View**
   > **Viewpoint**
   > **Axes**

   prompt: tripod/target as before
   respond: move cursor to quadrant B and left click
5. New model display with changed icon orientation.
6. Using either
   (a) DISPLAY–VPOINT-axes from the screen menu
   or (b) View–Set View–Viewpoint-Axes from the menu bar pick different circle quadrants in the target area.
7. Refer to Fig. 11.4 which displays the model with each quadrant selection.
8. Do not save the changes

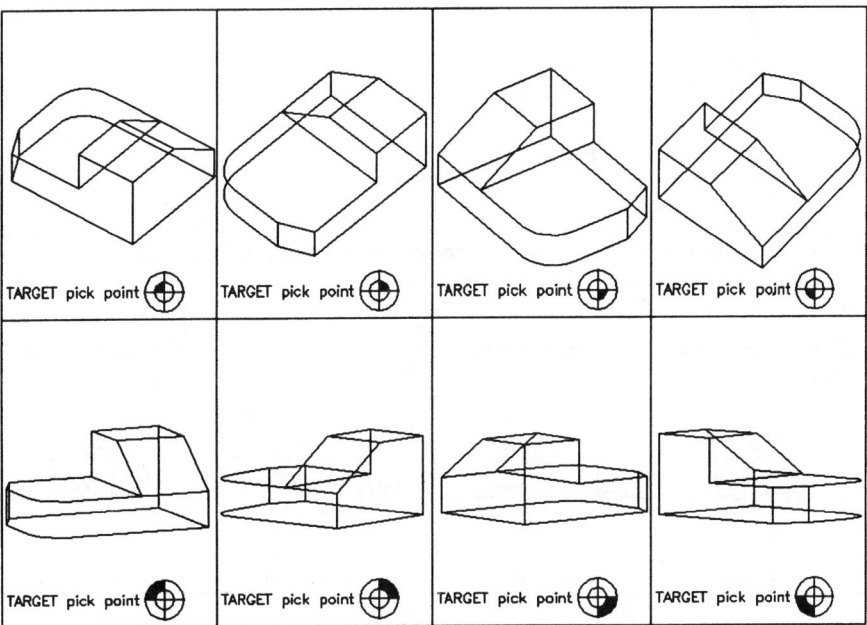

Fig. 11.4 VPOINT tripod axes/target dispalys using USEX1.

CIRCLE SELECTION

When using the target icon to select the viewpoint, the following should be remembered:
- inner quadrants – viewed from above
- outer quadrants – viewed from below.

VPOINT COORDINATE ENTRY

It is possible to use the VPOINT command with an *X, Y, Z* coordinate input to view a 3D model. The resultant display is similar to the rotate and axes options, but is generally used to obtain FRONT, TOP, SIDE, etc. views.

1. Open your **WORKDRG** drawing from the 3DPACK directory and:
   (a) make layer OUT current
   (b) freeze layers SECT, DIMENS and NOS
   (c) thaw layer TEXT
   (d) WCS current.
2. Select from the screen menu **DISPLAY**
   VPOINT:
   prompt: Rotate/<View point><.....>
   enter: **1,0,0** <R>
3. The screen will display a view of the model looking from the right side, the text RIGHT and V2 being 'readable'.
4. Using the VPOINT command, enter the following coordinates:

Coordinates	*View*
0,1,0	Rear
0,0,1	Top
−1,0,0	Left side
0,−1,0	Front
0,0,-1	Bottom
1,1,1	
−1,−1,−1	

5. The resultant views for the entered coordinates are given in Fig. 11.5.

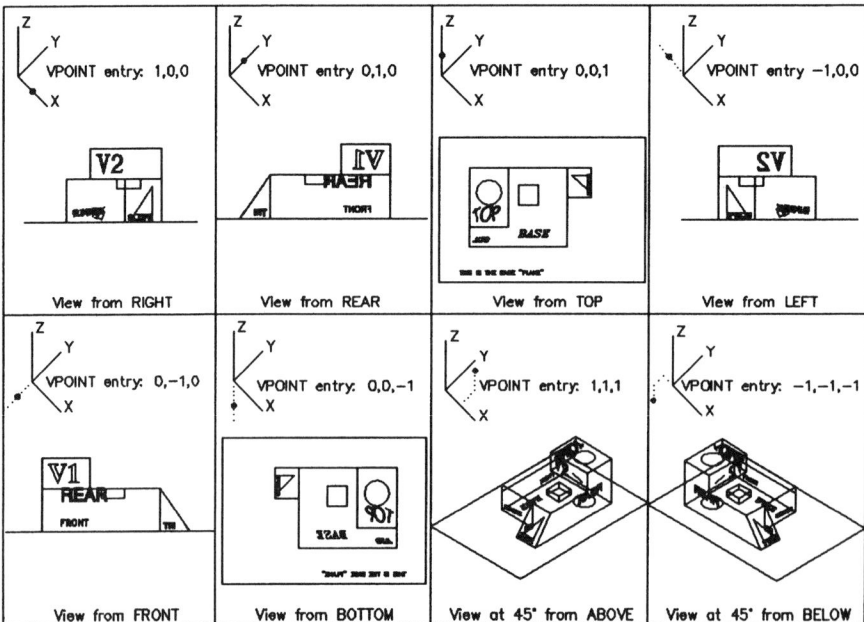

Fig. 11.5 VPOINT coordinate entry displays with WRKDRG.

THE ENTERED COORDINATES

To use the coordinate option of the VPOINT command, requires three coordinates to be entered from the keyboard. If you want to view the object from the right side, you must be standing on the positive portion of the *X*-axis looking towards the model. Hence the entry of 1,0,0. Any number could be entered, e.g. 5,0,0; 23,0,0, etc. I prefer to use the 1,0,0 entry.

   Similarly if you want to view the model from the front, you must be standing on the negative portion of the *Y*-axis, hence the entry of 0,–1,0.

Note

The coordinate entries are relative to the WCS. I'll leave you to investigate the coordinate entries with a named UCS setting. If you do use the coordinate option with a named UCS, observe the command prompt line as the command is executed. It makes interesting reading!

THE VPOINT PRESETS

The VPOINT command has preset values in dialogue box form and this can be selected:
(a) *from screen menu*   (b) *from menu bar*
   DISPLAY   View
   VPOINT: Set View
   View Pt   Viewpoint
                                    Presets...
Each selection results in the Viewpoint Presets dialogue box, which is displayed as 'two clocks'. By left clicking in the required quadrants of the clocks, then picking OK, the model will be displayed at the desired orientation. Personally, I find this option difficult to follow and very rarely use it. It is really the same as the VPOINT–Rotate option, which is easier to execute?

THE VPOINT PLAN OPTION

This option does what it says – it gives a plan (top) view of the model on the screen. The resultant plan view is, however, dependent on the UCS setting.
   1. Open your **ACTDRG** drawing of the hexagonal pyramid from the 3DPACK directory. Text and hatching displayed?
      (a) freeze the SECT layer if hatching added
      (b) make any other layer current
      (c) set to WCS – it probably is set?
      (d) make the 3D viewport active – it probably is (do you remember how?)
   2. Select from the screen menu **DISPLAY**
                              VPOINT:
                              plan

3. A plan (top) view of the model will be displayed.
4. Repeat the DISPLAY–VPOINT–plan selection in the other two viewports making each one active in turn.
5. The result is the same top view in the three viewports. This is due to the WCS setting. The plan option of the VPOINT command will always give the same view when the WCS is current.
6. With the right viewport active:
   (a) restore UCS FACE1
   (b) select from the screen menu DISPLAY–VPOINT–plan
   (c) no change in display?
7. From the menu bar select **View**
   Set View
   Plan View
   Current UCS
8. A plan view of the model is obtained. This is relative to the current UCS setting which is FACE1. The plan view is therefore equivalent to a front view.
9. With the top left viewport active:
   (a) restore UCS FACE2
   (b) select View–Set View–Plan View–Current UCS.
10. With the lower left viewport active:
    (a) make UCS SLOPE1 current
    (b) View–Set View–Plan View–Current UCS selection.
11. The resultant plan views of the model are now different (Fig. 11.6) due to the UCS setting. This can be very useful.
12. Do not save the changes.

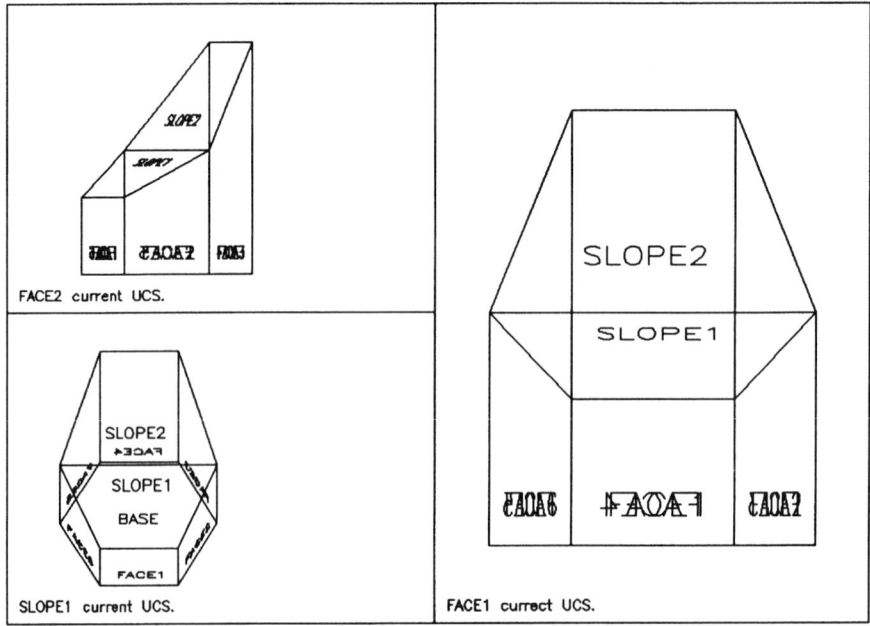

Fig. 11.6 ACTDRG with VPOINT – plan option using named UCSs.

TASK

1. WORKDRG

1. Open your WORKDRG from the 3DPACK directory.
2. The 3D model has a sloped face.
3. Can you view the model so that you are looking perpendicularly onto this sloped face. (Hint: one of the VPOINT options does it for you – but which one?)
4. Can you use a VPOINT option which will give the angle of the slope of this face to the horizontal. It should be 35.54°.
5. Do not save any changes.

2. EX11_2

1. Open the EX11_2 drawing from the 3DPACK directory.
2. 3D model displayed in red with green numbers in a four viewport configuration.
3. Can you use the VPOINT command to set each viewport, such that each number is viewed at 'right angles' – see Fig. 11.7. (Hint: think UCS.)
4. Save changes?

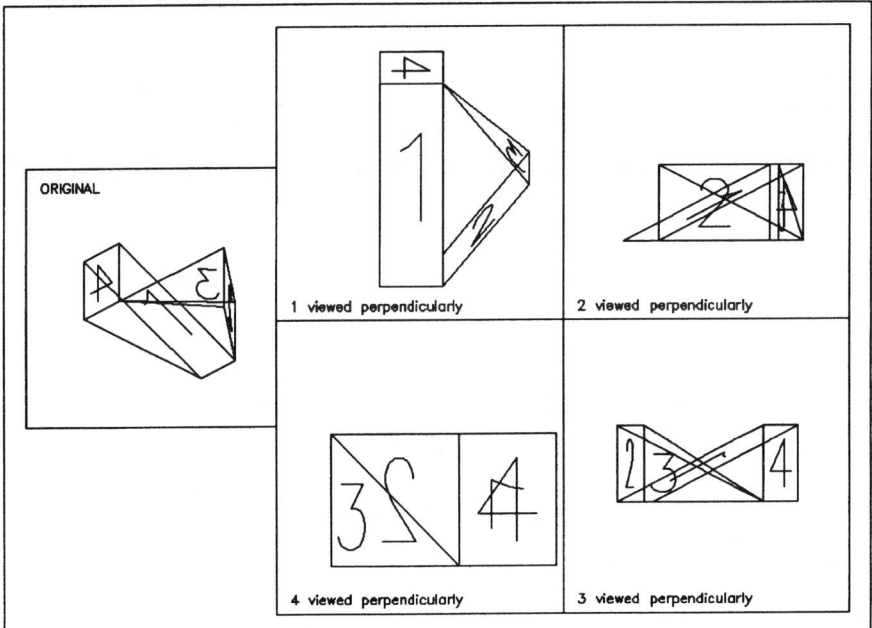

Fig. 11.7   Investigating VPOINT.

SUMMARY

1. VPOINT is the command which allows the user to view 3D models from different viewpoints.
2. The command can be activated from the screen menu, the menu bar or by direct keyboard entry.
3. The command has several options:
   (a) rotate – requires two angles to be entered
   (b) tripod/target – gives infinite viewpoints
   (c) coordinate entry – for front, side, top views
   (d) plan – for top views, but depends on the UCS setting.
   (e) presets – similar to the rotate option but is in dialogue box format.
4. The VPOINT command is generally used with multi-screen viewports, which will be discussed in the next chapter.
5. Personally I use:
   (a) the VPOINT–Rotate option with 315,30 as my angles. This gives a reasonable 'isometric' 3D view
   (b) the coordinate entry method to obtain the front (0,–1,0); right (1,0,0); top (0,0,1) views for orthographic layouts.

ACTIVITY

With all the exercises, examples and tasks in this chapter, I have not included any activites.

12

Multiscreen views – tiled viewports

Many of the drawing exercises encountered so far have been displayed in multiscreen mode, i.e. with more than one view of the model. No attempt was made with these early drawings to explain multiscreen drawings, but in this chapter we will investigate how they are set up.

Multiscreen drawings involve **viewports**, and AutoCAD allows the user to work with two types of viewport:
1. **Tiled** – which are fixed and cannot be altered by the user.
2. **Untiled** – which can be altered as required by the user.
Figure 12.1 shows both types of viewport. In this chapter we will only consider **tiled** viewports, and leave the **untiled** viewports to a later chapter.

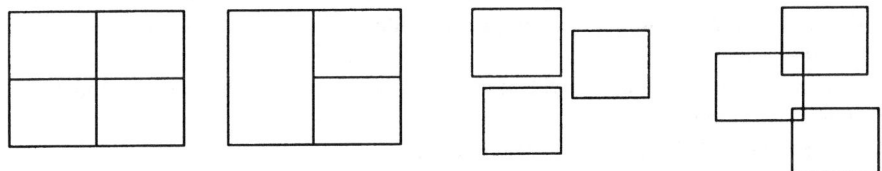

(a)   TILED viewports
      User cannot alter.

(b)   UNTILED viewports
      User can alter positions.

Fig. 12.1   TILED and UNTILED viewports.

WORKED EXAMPLE

As usual we will investigate **tiled** viewports with an example. The example is quite involved, but it will be worth the time and effort spent.

1. Open drawing EX12_1 from the 3DPACK directory.
2. The screen displays a red 3D wire-frame model on a black 'base'.
3. From the screen menu select **SETTINGS**

   next

   VPORTS:

   prompt: Save/Restore/Delete/Join/SIngle/?/2/<3>/4

   enter: **4 <R>**
4. The screen will display four viewports each with the same view of the wire-frame model.
5. Using the VPOINT command, enter the following coordinates:

   (a) top left   VPOINT 0,–1,0   FRONT view

   (b) top right   VPOINT 1,0,0   RIGHT side view

   (c) bottom left   VPOINT 0,0,1   TOP (plan) view

   (d) bottom right   leave as is   3D view

   Note: remember that a viewport is made **active** by moving the mouse into it, then left clicking.
6. The model is displayed as in orthogonal first angle projection.
7. Do your views 'line up with each other'?

CENTRING VIEWPORTS

When the VPOINT command has been used in several viewports, the resultant views of the model generally do not 'line up', i.e. the side view may be a different 'size' from the front view.

One method of lining up individual viewports is to use the ZOOM command with the CENTRE option.

1. Make the FRONT (top left) viewport active.
2. From the screen menu select **AutoCAD**

   DISPLAY

   ZOOM

   Centre

   prompt: Center point

   enter: **190,130,20 <R>**

   prompt: Magnification or Height<?>

   enter: **300 <R>**
3. Repeat the ZOOM–Centre option in the top right and bottom left viewports, entering the **SAME** centre point (190,130,20) and magnification (300).
4. Leave the 3D viewport as it is.
5. Your screen display should be similar to Fig. 12.2.

ZOOM–CENTRE

This option will centre the viewport about the entered point. The user is then prompted for a magnification factor. By entering the 'model centre point', the viewport will be 'zoomed' about this point by the factor entered for the magnification. A large value for magnification gives a smaller model. The value actually entered may require some 'trial and error' by the user, but my values should give a reasonable result.

SAVING A VIEWPORT CONFIGURATION

When different viewport configurations are used, it is useful to save them for future recall (similar to the UCS). The procedure is very straightforward.

At the command line enter **VPORTS** <R> (or pick from screen menu)

prompt: Save/Restore/...

enter: **S** <R>

prompt: ?/Name for new viewport configuration

enter: **4centred** <R>

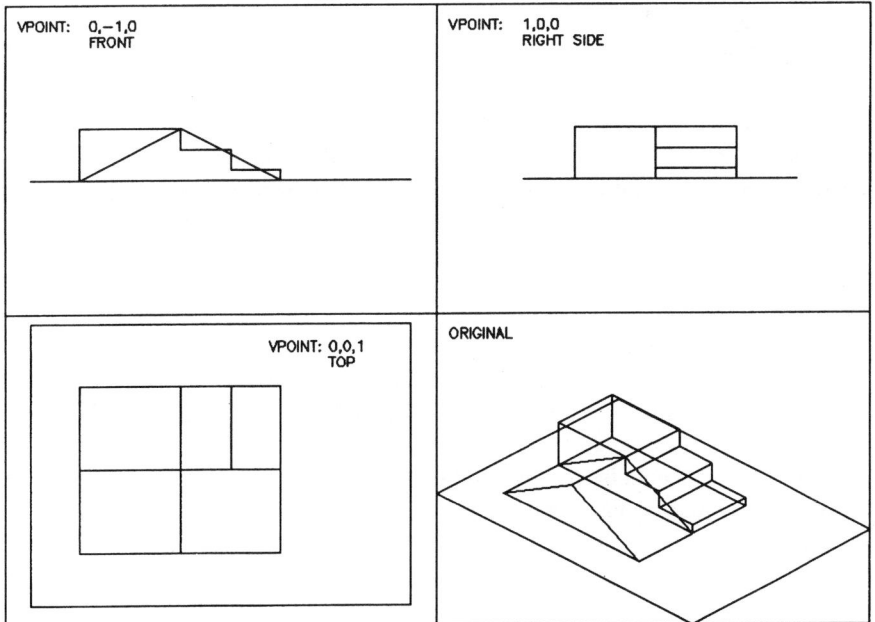

Fig. 12.2   EX_12_1 in four-viewport configuration.

ANOTHER VIEWPORT CONFIGURATION

1. Make the original 3D viewport active.
2. From the screen menu select **SETTINGS**
   next
   VPORTS
   Single
3. The screen will display a single 3D viewport.
4. Activate the VPORTS command again and select **3**
   prompt: Horizontal/Vertical...
   enter: L <R>
5. The screen will display a three-viewport configuration with the large viewport on the left.
6. Change the VPOINT to:
   (a) top right: 1,0,0 FRONT
   (b) bottom right: 0,0,1 TOP
7. Centre the two smaller viewports using the same values as before, i.e. ZOOM–Centre option at 190,130,20 with 300 magnification.

ADDING OBJECTS TO THE MODEL

1. Make OBJECTS layer (blue) current.
2. Restore the named UCS SLOPE.
3. With the 3D viewport active, use the LINE command to draw:
   from: 20,20
   to: @50,0
   to: @0,60
   to: @−50,0
   to: close.
4. A blue rectangular shape is added to the model in all viewports.
5. Restore UCS TOP and draw a circle, centred at 50,40 with a radius of 15.

TASK

Try the following additions.
1. Draw two other circles of radius 15 at the centre of the two horizontal surfaces.
2. Draw a rectangular shape similar to the one already drawn on the other sloped surface.
   Hint: new UCSs are needed!
   The final result should be as shown in Fig. 12.3.
   Save this viewport configuration as 3left.

RESTORING THE FIRST VIEWPORT CONFIGURATION

1. Activate the VPORTS then R to the prompt and
   prompt: ?/Name of viewport to restore
   enter: **4centred** <R>
2. The four-viewport configuration will be displayed, with the blue objects added.
   Now save your exercise as **EX12_1A** in the 3DPACK directory.

Fig. 12.3 EX12_1 in three-viewport configuration.

SUMMARY

1. Tiled viewports allow multi-screen views of 3D models.
2. The viewport command is activated from:
   (a) the screen menu with **SETTINGS–next–VPORTS**
   (b) the keyboard with **VPORTS**.
3. Multi-viewports are generally used with the VPOINT command to set up FRONT,
   TOP and SIDE views.
4. Viewport configurations can be saved for future recall.
5. Modifications to the active viewport are displayed in the other viewports.

ACTIVITY

This is a relatively easy activity, but is quite long. It will test some of the work in previous chapters.
1. Open the **ACT_5** drawing from the 3DPACK directory.
2. Screen displays a pyramid type model.

Viewport set-up

1. Create a four-viewport configuration to show a front, top and side view as well as the 3D view.
2. Centre the viewports about the point 190,130,80 at 300 times magnification.

Using saved UCSs

1. Three named UCSs have been set.
2. Make OBJECTS the current layer.
3. Draw a 20 radius circle, centre at 25,40 with each named UCS current – Fig. 12.4.
4. Save the configuration as config1.
5. Return to WCS.

Second viewport configuration

1. With the 3D viewport active, set a single viewport configuration.
2. Set up a three-viewport configuration with:
   (a) the large viewport in its original 3D view
   (b) the other viewports set to VPOINT–Rotate with angles of:
       (i) 20 and 20 and
       (ii) 20 and −20.

Object addition

1. Add a circle, radius 20 to the centre of the top face – Fig. 12.5. Is a new UCS needed?
2. Save the configuration as config2.
3. Restore viewport configuration config1 – blue circle added?
4. Save your work.
   This completes the viewport activity.

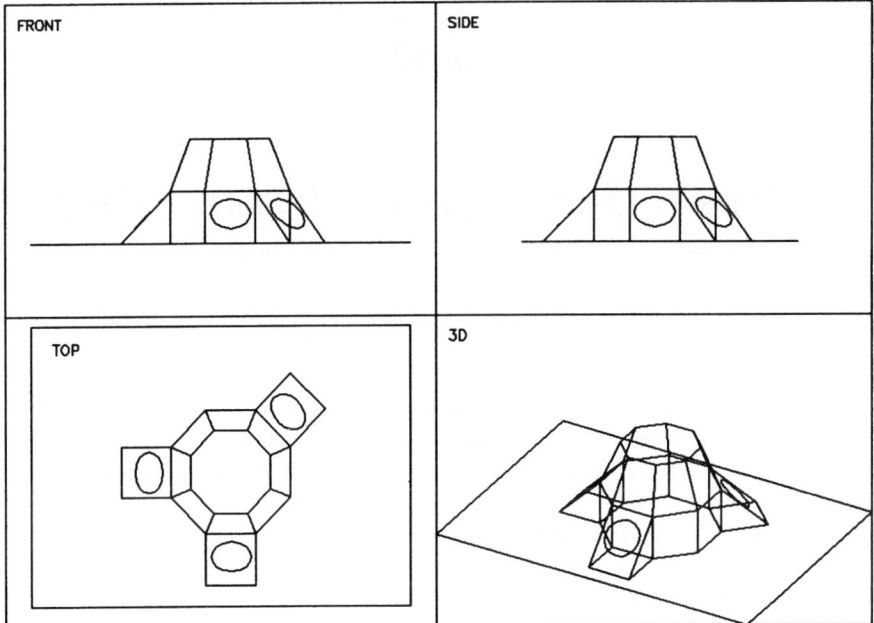

Fig. 12.4   ACT_5 in four-viewport configuration with circles added.

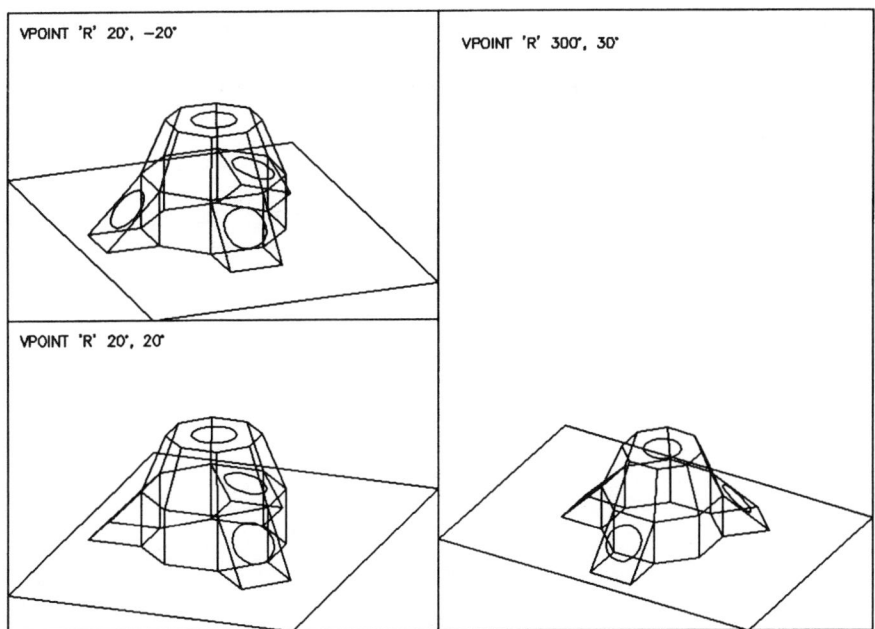

Fig. 12.5   ACT_5 in three-viewport configuration with circle added to top surface.

13

Three-dimensional surfaces

AutoCAD has several types of 3D surfaces which can be added to models. These surfaces can convert a wire-frame model into a surface model. Surface models are more 'realistic' looking than wire-frame models as the ambiguity effect of the wire-frame model is removed.

The surfaces which are available with AutoCAD are:

(a) edge surfaces
(b) revolved surfaces
(c) ruled surfaces
(d) tabulated surfaces
(e) 3D faces
(f) 3D meshes
(g) polyfaces
(h) 3D polylines.

Each surface will be considered as a separate chapter, although not necessarily in the order listed above.

The exercises will use many of the commands previously used.

14

Three-dimensional faces

A 3D face is a three- or four-sided shape added to any flat surface, and is a very simple way of converting a wire-frame model into a surface model.

EXERCISE 1

1. Open exercise **EX14_1** from the 3DPACK directory.
2. The screen will display a red pentagonal pyramid on a black base in a four-viewport configuration as shown in Fig. 14.1.
3. With the 3D viewport active (it should already be) enter **HIDE** <R>.
4. Nothing happens, as the pyramid is a wire-frame model with no surfaces on the sloping sides.

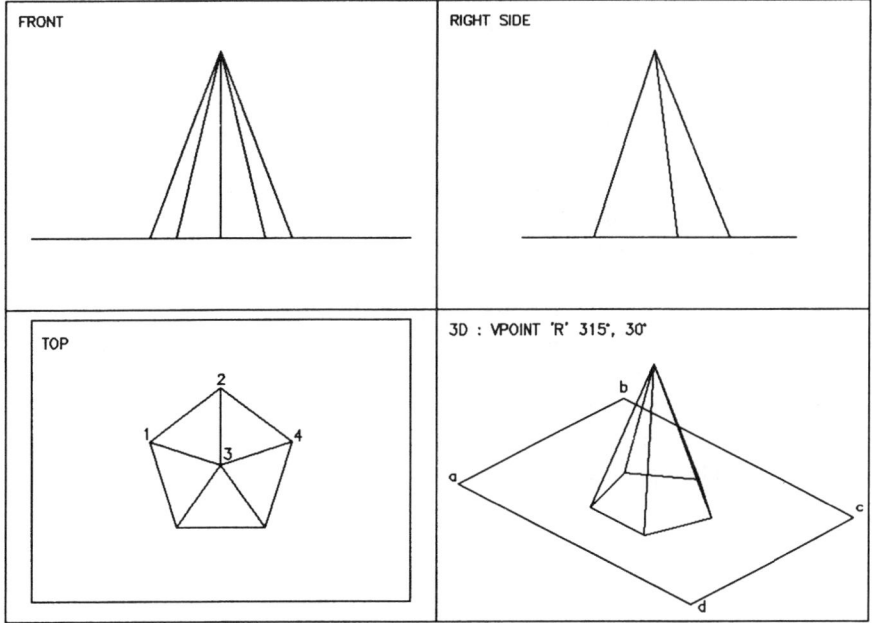

Fig. 14.1 3DFACE exercise EX14_1.

5. Activate the layer control dialogue box, and note:
   (a) layers FACE1–FACE5 are new with different colours
   (b) layers FACE1–FACE3 are frozen.
6. Thaw layers FACE1–FACE3.
7. The wire-frame model will be displayed with yellow, green and blue lines.
8. Making each viewport active, enter HIDE <R>. Are the hidden lines removed? *Note*: the phrase 'hidden lines' refer to any entities of a model which are 'behind the field of sight'.
9. With the 3D viewport active, change the viewpoint with the rotate option, entering angle values of 30 and 30. Now enter HIDE.
10. At the current viewpoint, you should be able to 'see' into the pyramid, as two of the sloped sides do not have surfaces.
11. Making each viewport active, enter **SHADE** <R> and you should be impressed with the coloured model that results.
12. From the screen menu select **DISPLAY**
   REGNALL
13. This removes the SHADE and HIDE effects in all viewports.

ADDING 3D FACES

We will now add 3D faces to the two remaining slopes of the pyramid.
1. Make the TOP viewport active (bottom left).
2. Make FACE4 the current layer and freeze layers FACE1–FACE3.
3. Set a running OSNAP to END.
4. From the screen menu select **AutoCAD**
   SURFACES
   3DFACE:
   prompt: First point and pick pt 1 (refer to Fig. 14.1)
   prompt: Second point and pick pt 2
   prompt: Third point and pick pt 3
   prompt: Fourth point and pick pt 1
   prompt: Third point and <RETURN> or right click
5. A magenta outline will be drawn about the relevant sloped surface in all viewports.
6. Make FACE5 current and freeze FACE4.
7. At the command line enter **3DFACE** <R> and:
   prompt: First point and pick pt 2
   prompt: Second point and pick pt 4
   prompt: Third point and pick pt 3
   prompt: Fourth point and pick pt 2
   prompt: Third point and right click
8. A cyan outline will be displayed in all viewports.
9. Now:
   (a) make layer OUT current
   (b) thaw layers FACE1–FACE 4
   (c) enter HIDE <R> in each viewport.

10. The pentagonal pyramid will be displayed as a complete surface model in all viewports. There are no hidden lines. *Note*: the above statement is not really true. The pyramid has a base surface, and this has not been 'surfaced'.
11. Enter **REGENALL** <R> to remove the hide effect in all viewports.
12. Now SHADE each viewport – an interesting effect?
13. In the 3D viewport try the SHADE command with some other VPOINT settings, e.g.
    (a) VPOINT 'R' 30, 80
    (b) VPOINT 'R' 150,70
14. Return the 3D viewport to its original 315,30 angle values.
15. REGENALL – keyboard or screen menu.
16. Make OUT the current layer.
17. Using the 3DFACE command:
    prompt: first point and pick pt a (see Fig. 14.1)
    prompt: second point and pick pt b
    prompt: third point and pick pt c
    prompt: fourth point and pick pt d
    prompt: Third point and <RETURN>
18. SHADE to give the coloured pyramid on a red base.
19. REGENALL all and save your work.

ACTIVITY

This is a fairly easy activity and involves a hexagonal pyramid on a hexagonal prism.
1. Open the drawing **ACT-6** from the 3DPACK directory.
2. The screen displays a four viewport configuration of the wire-frame model – Fig. 14.2.
3. With the 3D viewport active, enter HIDE then SHADE – nothing different as the model has no faces.
4. Investigate the layer control dialogue box and note that there are several new layers – FACE1A, FACE1B, etc. These layers are for adding 3Dfaces to the various 'surfaces' and A layers are for facing the vertical sides of the prism B layers are for facing the sloped sides of the pyramid.
5. Thaw the six frozen layers, then HIDE each viewport – all lines 'behind' the faces are not shown.
6. Try SHADE in each viewport then REGENALL.
7. Now complete 3D 'facing' the model. FACE4 layer has been added for you, so four(?) new layers are required. Use colours for these new layers.
8. When the model has been faced, HIDE and SHADE.
9. Exit when you've had enough – save?

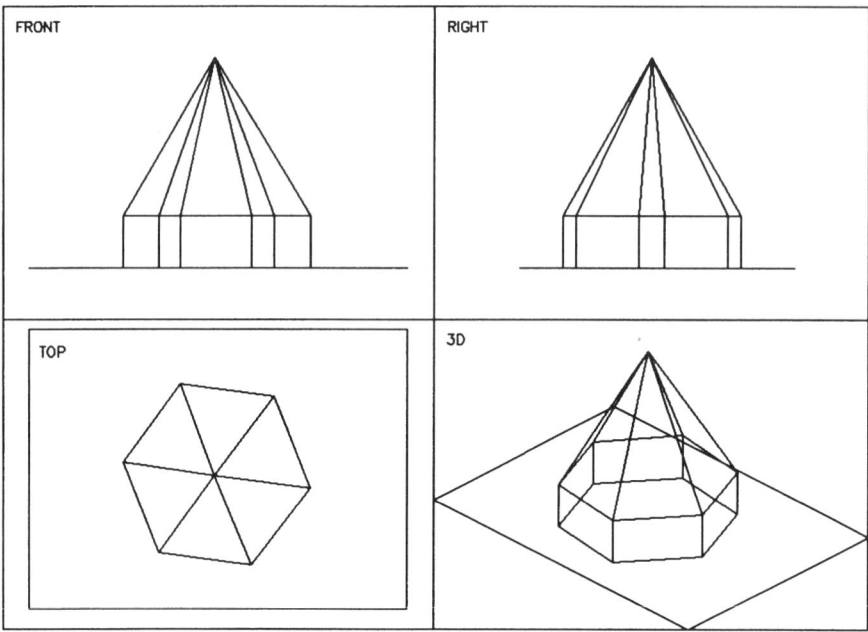

Fig. 14.2   Model for activity 6.

3DFACE: THE INVISIBLE EDGE

Only three- and four-sided figures can have 3Dfaces added. When a complicated shape is required to be faced, the command can still be used and the user can use the **Invisible Edge** option.

1. Open drawing **EX14_2** from the 3DPACK directory.
2. Screen displays two similar red shapes.
3. The shapes have had 3Dfaces added:
   (a) the left shape without the invisible option
   (b) the right shape with the invisible option used.
4. Using the layer control dialogue box, thaw layer FACE1 and make it current.
5. Yellow faces are displayed?
6. Select from the screen menu **SURFACES**
   3DFACES:
   ShowEdge
   prompt: Invisible edges will be SHOWN after next Regeneration
   respond: Cancel the command (CTRL C) and enter **REGEN** <R>
7. The invisible edges are displayed in the right shape.
8. Thaw layer SECT and hatching will be added to the two shapes.
9. With the 3DFACE command, select HideEdge and:
   prompt: Invisible edges will be HIDDEN after next Regeneration
   respond: Cancel the command (CTRL C) and enter **REGEN** <R>
10. The invisible edges are removed in the right view.
11. Compare the hatching in each view – it is the same?

USING THE INVISIBLE EDGE OPTION

1. Still with EX14_2 and layer FACE1 current, thaw layer TRIAL.
2. Two numbered shapes are displayed in black.
3. Running OSNAP to END?
4. Using the 3DFACE command and the left shape:
   prompt: first point and pick pt 1
   prompt: second point and pick pt 2
   prompt: third point and pick pt 3
   prompt: fourth point and pick pt 4
   prompt: third point and pick pt 5
   prompt: fourth point and pick pt 6
   prompt: third point and pick pt 7
   prompt: fourth point and pick pt 8
   prompt: third point and <RETURN>
5. A yellow 3Dface is added to the shape, with an 'edge' between points 3 and 4 and points 5 and 6.
6. With the right shape, use the 3DFACE command and:
   prompt: first point and pick pt 1
   prompt: second point and pick pt 2
   prompt: third point
   respond: pick **Invisible** from screen menu then pick pt 3
   prompt: fourth point and pick pt 4
   prompt: third point
   respond: pick **Invisible** from the screen menu then pick pt 5
   prompt: fourth point and pick pt 6
   prompt: third point and pick pt 7
   prompt: fourth point and pick pt 8
   prompt: third point and <RETURN>
7. The two trial shapes have been faced, the right shape having the invisible edge option.
8. *Task*: try hatching these yellow 3Dfaces!
This completes the 3DFACE section.

SUMMARY

1. The 3DFACE command allows the user to add surfaces to a wire-frame model.
2. The command can only be used with three- or four-sided figures.
3. When used, a wire-frame model is converted into a surface model.
4. The HIDE command used with a surface model will 'hide hidden lines', i.e. lines and surfaces 'behind' the viewpoint.
5. 3Dfaces should always be added *on their own layer*, i.e. a separate layer for each face.
6. Coloured 3Dfaces give a useful effect with the SHADE command.
7. A 3Dface is a *single entity*.
8. The commands:
   (a) REGEN, removes the HIDE and SHADE effect from the active viewport.
   (b) REGENALL, removes the HIDE and SHADE effect from all viewports.

15

Three-dimensional mesh

A 3D mesh can be considered as a type of rectangular matrix having length, width and height. AutoCAD refers to the length and width as the mesh M and N sizes. Every (M,N) point on the mesh has an x,y,z coordinate which is called the *vertex* of the mesh. When a mesh has been created, it can be extensively edited and there are several 'spline fitting' curve techniques available to 'smooth' the mesh.

1. Open drawing **EX15_1** from the 3DPACK directory.
2. The screen displays an 8 × 6 3Dmesh in a four-viewport configuration similar to that shown in Fig. 15.1 at different VPOINT 'R' values.
3. Study the 3Dmesh display. I have added coloured donuts at the four mesh corners, merely to give an idea of the mesh orientation. The mesh represents a 'landscape' with a 'depression' in it. Note that one of the mesh vertices is wrong, and we want to edit this vertex.

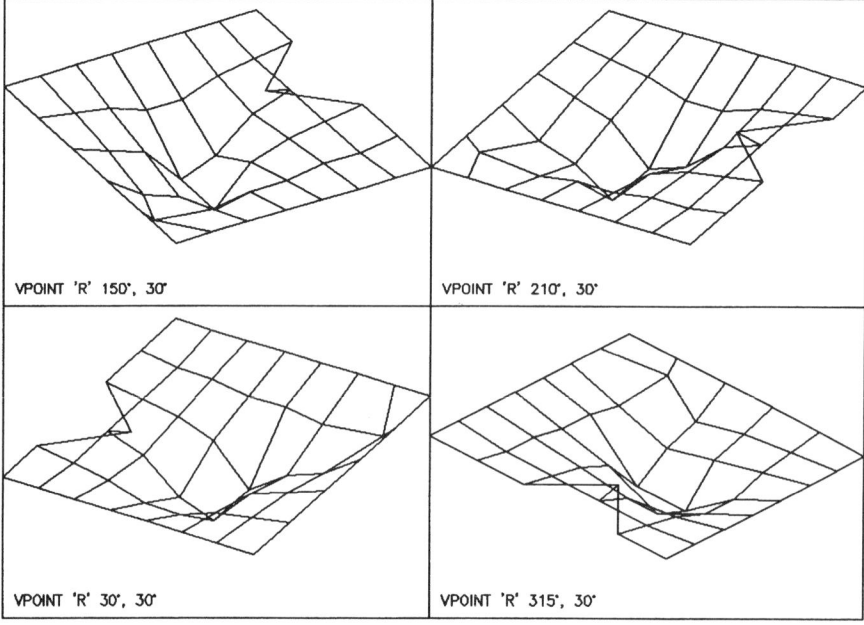

Fig. 15.1 EX15_1 original 3Dmesh.

4. A 3Dmesh is a single entity and is made of polylines. The PEDIT command is therefore available to us on it.

5. Ensure the lower right viewport is active, then select from the menu bar **Modify**

                                                                     PolyEdit

   prompt: Select objects

   respond: **pick any point on the 3Dmesh then <R>**

   prompt: the polyedit options are displayed both in the screen menu area and at the command line

   respond: **pick EdVrtx from the screen menu**

   and   (a)   new prompt – Vertex(0,0) Next/Previous…

             (b)   a white cross appears at vertex (0,0) – at the cyan donut.

             (c)   we want to move the cross to the vertex to be edited

   enter: **N <R>** – the Next option

   prompt: Vertex(0,1) Next/Previous…

   enter: **N <R>** – the next option again

   prompt: Vertex(0,2) Next/Previous…

   enter: **N <R>** – the next option again

   prompt: Vertex(0,3) Next/Previous…

This is the vertex to be edited

   enter: **M <R>** – the move option of EdVrtx

   prompt: Enter new location

   enter: **@0,–50,0 <R>** – I originally created the 3Dmesh with the vertex (0,3) having a y-coordinate which was 50 units 'too big'.

   prompt: Vertex(0,3) Next/Previous…

   enter: **X <R>** – to exit the EdVrtx options

   prompt: Edit vertex/Smooth surface…

   enter: **S <R>** – smooth surface option

   prompt: Generation segment…

   then   (a)   redraws mesh as a smooth surface

             (b)   returns the prompt line

   enter: **X <R>** – to end the PolyEdit command.

We have thus edited the 3Dmesh and applied a smooth surface to it. The effect is quite impressive, as you can now 'see' the depression in the landscape – Fig. 15.2.

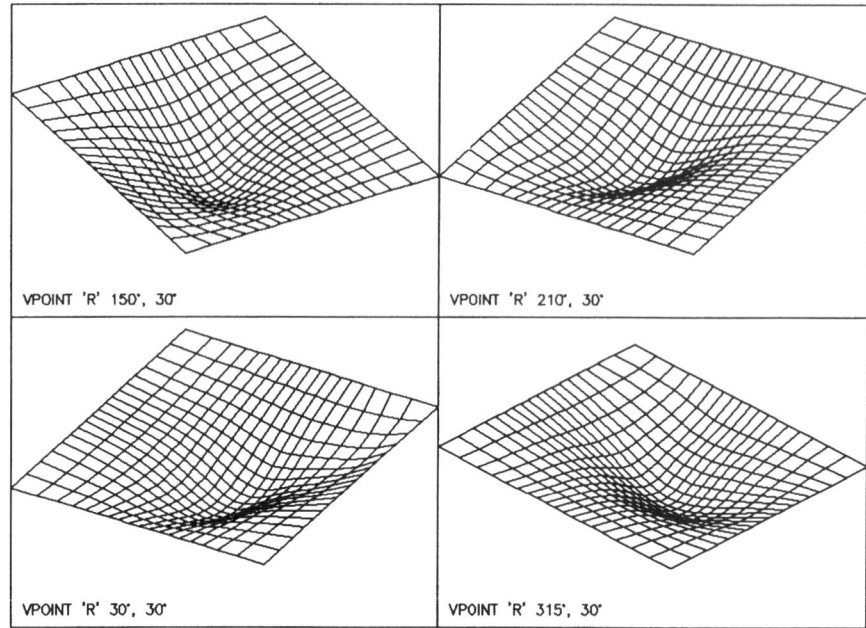

Fig. 15.2 EX15_1 original 3Dmesh.

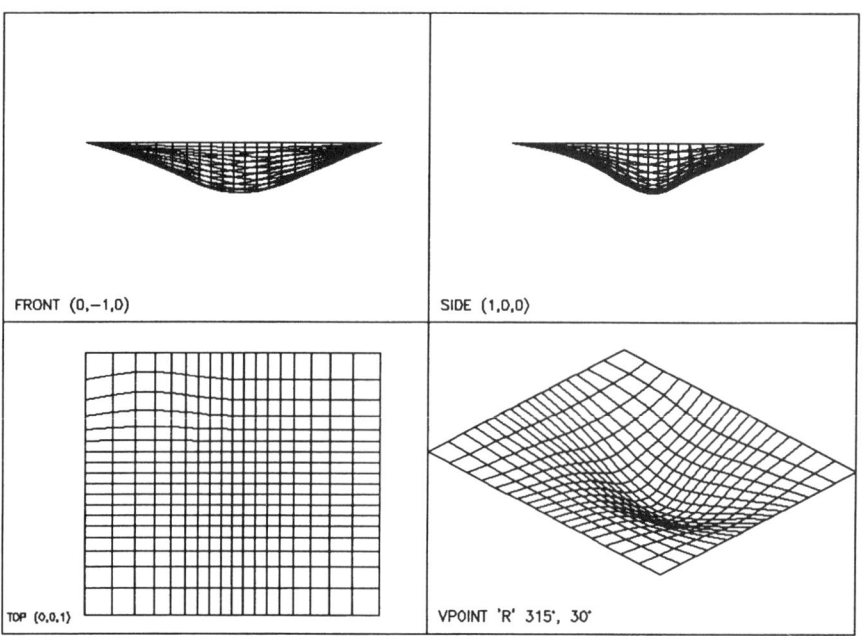

Fig. 15.3 EX15_1 after EdVrtx–Move option and Smooth option.

SMOOTHED SURFACES

AutoCAD allows the user three types of smoothing techniques which can be added to mesh type surfaces, these being:
1. Quadratic fit polymesh.
2. Cubic fit polymesh.
3. Bezier fit polymesh.
These 'fit polymeshes' have several variables which can be altered by the user, and include the following:

- *SPLFRAME*
  Controls whether the polymesh is added to the surface:
  0 (off) – polymesh is added (default setting)
  1 (On) – no polymesh added, i.e. original 3Dmesh only.

- *SPLSEGS*
  Controls the number of segments in the polymesh – the default value is 8.

- *SURFTYPE*
  Controls what type on polymesh is to be added
  5 – quadratic (default)
  6 – cubic
  8 – Bezier.

Unless you are 'specializing' in meshes, I would suggest that the default values are suitable for your applications. You can investigate the types of curves if you want, as they are easy to activate from the screen menu when the PolyEdit command is used. *Note*: the PolyEdit command from the menu bar, is the same as the PEDIT command from the screen menu.

TASK

1. Change the viewpoint to give a top, front and side view as shown in Fig. 15.3. Use a ZOOM Centre about 150,150,–20 at 250 magnification to 'centre' the views.
2. Try HIDE and SHADE in each viewport – useful?

ACTIVITY

Creating a 3Dmesh is usually very boring, as it requires the user to enter the x,y,z coordinates of every vertex in the mesh. I have got one for you to attempt, which should keep you busy for a few minutes.

1. Open the drawing **ACT_7** from the 3DPACK directory.
2. The screen displays a four-viewport configuration with nothing. The viewports have been set up and centred for the activity.
3. With the lower-right viewport active and layer MESH current (they should be), use the **SURFACES–3DMESH** command and enter the following:

   mesh M size: 5
   mesh N size: 5

Vertex	Coords
0,0	40,40,0
0,1	90,40,0
0,2	140,40,0
0,3	190,40,−80
0,4	240,40,0
1,0	40,70,0
1,1	90,70,−5
1,2	140,70,−10
1,3	190,70,−15
1,4	240,70,−20
2,0	40,100,0
2,1	90,100,−15
2,2	140,100,−25
2,3	190,100,−40
2,4	240,100,−60
3,0	40,130,50
3,1	90,130,−5
3,2	140,130,−10
3,3	190,130,−15
3,4	240,130,−20
4,0	40,160,0
4,1	90,160,0
4,2	140,160,0
4,3	190,160,0
4,4	240,160,0

4. When the 3Dmesh is complete, two of the vertices will be displayed wrongly, due to errors in the coordinate input. Using the PolyEdit command, use the Move option of EdVrtx to change the following vertices:
   (a) vertex(0,3) to 190,40,0
   (b) vertex(3,0) to 40,130,0
5. Smooth curve the surface. Figure 15.4 displays the 3Dmesh as a cubic fit polymesh.
6. Try HIDE and SHADE.

This completes the 3Dmesh exercises.

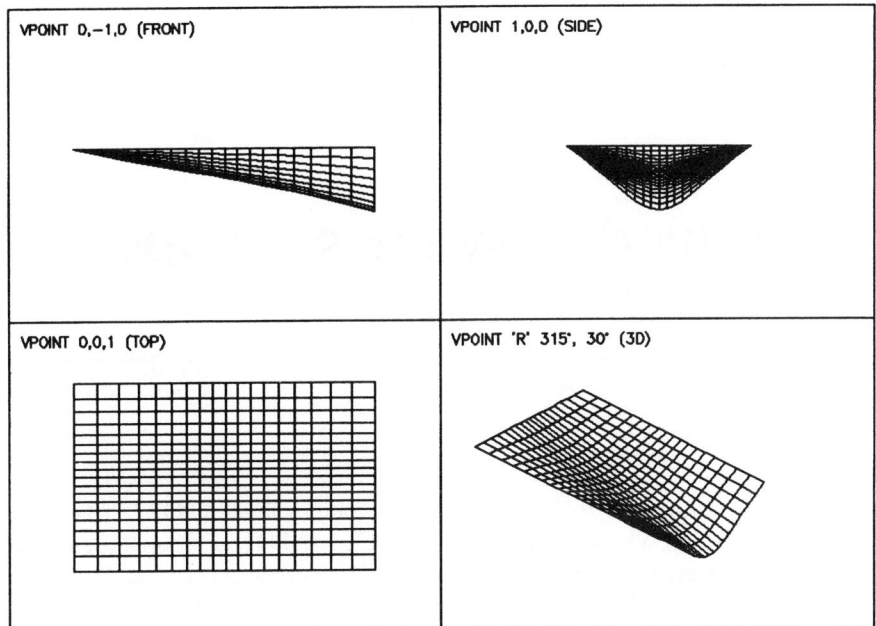

Fig.15.4   ACT_7 3DMESH as a cubic smooth surface.

16

Untiled viewports – Paper space

For the last few chapter we have been using tiled viewports with the VPORTS command. These viewports have allowed us to view a 3D model at different viewpoints, and we have been able to produce front, top, side and 3D views on the one screen. While tiled viewports are very useful, they have several limitations which include:

    (a) the viewport layout is fixed

    (b) additional viewports cannot be added

    (c) dimensions in 'individual' viewports is not possible

    (d) only the active viewport can be plotted.

AutoCAD has an added feature which allows the user to create viewport configurations which overcome the tiled viewport limitations. These are *untiled* viewports and introduce the paper and model space concept.

    To demonstrate untiled viewports we will use several examples.

EXAMPLE 1

  1. Open your **WORKDRG** drawing from the 3DPACK directory – I hope you still have it. The drawing should be displayed:

    (a) as a single viewport at viewpoint 315,30

    (b) with hatching?

    (c) with a yellow viewport border

    (d) inside a black border.

  2. Make layer 0 current, freeze layer SECT. WCS current.

  3. Use the DTEXT command to enter the following text:

    (a) start point 30,30

    (b) height 10 and rotation angle 0

    (c) text: This is MODEL SPACE TEXT.

  4. Note the text orientation.

  5. Draw a circle, centre 380,120 with radius 20.

Example 1   79

6. From the menu bar select **View**
                                Paper Space
   and note: (a)   new paper space icon – Fig.16.1
             (b)   the letter P in the Status bar.
7. With the DTEXT command enter the following text:
   (a)   start point 30,30
   (b)   height 10, rotation angle 0
   (c)   text: This is PAPER SPACE TEXT.
8. Draw a circle with centre 380,120 and radius 20 – Fig. 16.2.
9. Compare the text and circle orientation.
10. With the ERASE command, try and erase any entity of the wire-frame model. You cannot!
11. From the menu bar select **View**
                                Model Space
    The P will be removed from the status bar and the WCS icon will return.
12. Try and erase the text/circle created in Paper space, and you will find that you cannot.
13. Do not save these changes.

Fig. 16.1   Paper space icon.

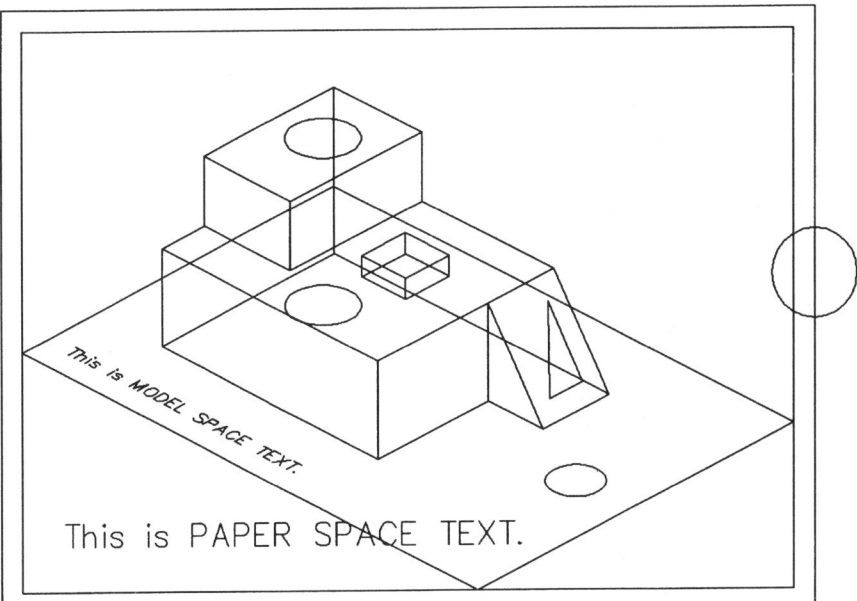

Fig. 16.2   WORKDRG in Paper space mode.

MODEL/PAPER SPACE

Model space and Paper space are two different 'drawing modes' available on the one screen. In a way they can be considered similar to layers, and drawings can be created on each 'layer'. The user can also 'toggle' between the two modes.

Model space

This is the default mode used to create the actual model in 2D or 3D. In this mode, multi-screen viewports are created with the VPORTS command.

Paper space

Paper space is a drawing mode which has several advantages to the user:
1. Multiple viewports are created with MVIEW.
2. Additional viewports can be added.
3. Viewport positions can be altered.
4. All viewports can be plotted at the one time.

The two modes can be activated:
1. From the menu bar with View then Model space or Paper space.
2. From the screen menu with MVIEW then MSPACE or PSPACE.
3. From the keyboard by entering PS <R> and MS <R>. This is my preference.

We will now proceed with an exercise which will create multiple viewports in Paper space.

EXAMPLE 2

1. Open drawing **EX16_1** from the 3DPACK directory.
2. The screen will display a black border with the Paper space icon.
3. Make VP the current layer.
4. Select from the menu bar **View**
                        Mview
                        4 Viewports
    prompt: Fit/<First point>
    enter: **10,10 <R>**
    prompt: Second point
    enter: **370,260 <R>**
5. The screen will display four yellow viewports with nothing in them.
6. Enter Model space with **MS <R>**.
7. Change the viewpoint with the VPOINT command to:
    upper left: VPOINT 0,–1,0 – front
    upper right: VPOINT 1,0,0 – right side
    bottom left: VPOINT 0,0,1 – top
    bottom right: VPOINT 'R' 315, 30

8. Now thaw the layer MODEL to display a square pyramid with coloured sides.
9. Zoom–centre about 100,100,90 at 250 magnification in all four viewports.
10. With the TEXT layer current and the 3D viewport active, use the DTEXT command to enter the following:
    (a) centred at the point 100,20
    (b) height 10, rotation 0.
    (c) text: PYRAMID.
11. The text will be added in all four viewports.

ALTERING THE VIEWPORT CONFIGURATION

1. Ensure the 3D viewport is active.
2. Enter Paper space with **PS** <R>
3. From the menu bar select **Modify**
    Stretch
    prompt: First corner and enter **375,5** <R>
    prompt: Other corner and enter **350,265** <R>
    prompt: 2 found
    then: Select objects
    enter: **<RETURN>**
    prompt: Base point and enter **370,260** <R>
    prompt: Second point and enter **330,260** <R>
4. Repeat the Modify–Stretch selection and enter:
    (a) first corner: 5,5
    (b) other corner: 20,265
    (c) base point: 10,10
    (d) Other point: @50,0
5. Using the MOVE command with the window option, window the four yellow viewports then <RETURN>
    prompt: Base point and enter 330,10
    prompt: Second point and enter @40,0
6. The four yellow viewports will be reduced in size and moved to the right of the drawing screen.

CREATING ADDITIONAL VIEWPORTS

1. Still in Paper space, make VP the current layer.
2. Select from the menu bar **View**
                                    Mview
                                    Create Viewport
   prompt: …/<First point> and enter **10,20** <R>
   prompt: Other corner and enter **100,135** <R>
3. A new viewport will be displayed with the 3D model and text.
4. Create another viewport with:
   (a)   first corner at 20,150
   (b)   other corner at 80,250.
5. Enter Model space (MS <R>) and:
   (a)   with the new lower viewpoint active, change to VPOINT 'R' 315 and −30.
   (b)   with the new upper viewport active, change to VPOINT −1,0,0.
6. ZOOM Centre about 100,100,90 at 250 magnification in the two new viewports.
7. Enter Paper space and make 0 the current layer.
8. Enter text, centred on 230,200 at a height of 6 with 0 rotation the text being:
   PYRAMID
   CREATED IN
   MODEL SPACE
   VIEWPORTS IN
   PAPER SPACE
9. The result should be as shown in Fig. 16.3.
10. Save if you want then proceed to the next example.

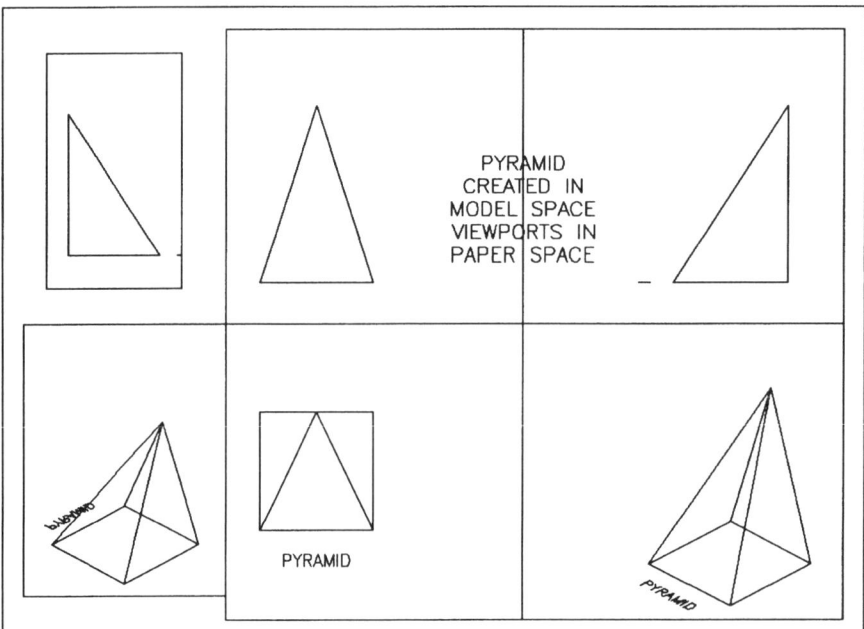

Fig. 16.3   EX16_1 model/Paper space exercise.

Example 3   83

EXAMPLE 3

1. Open the drawing **EX16_2** from the 3DPACK directory.
2. Screen displays a four viewport configuration of WORKDRG in Paper space.
3. In Paper space with layer 0 current, draw a circle at centre 375,5 with radius 5.
4. Rectangular ARRAY this circle:
   (a) for two rows and two columns
   (b) row distance 260
   (c) column distance −370.
5. Enter Model space (MS <R>) with lower right viewport active, repeat steps 3 and 4 using the same values.
6. Note the array differences in Model and Paper space.
7. In Paper space (PS <R>), select the MOVE command and:
   prompt: Select objects
   respond: **pick the yellow upper border of the upper left viewport then <RETURN>**
   prompt: Base point
   respond: **OSNAP END and pick the top left corner of the selected viewport**
   prompt: Second point
   respond: **OSNAP END and pick the top left corner of the black border**
8. Similarly move the other three viewports to the corner of the black border.
9. *Task*: create a new viewport in the centre of the screen at your own viewpoint – Fig. 16.4.
10. Do not quit this drawing.

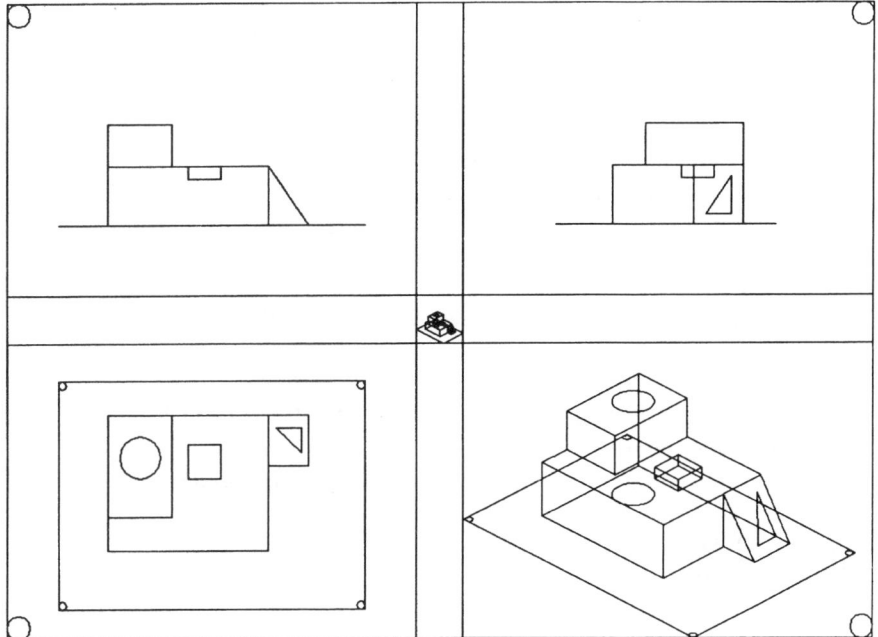

Fig. 16.4   EX16_2 after modifications.

ZOOMING IN MODEL/PAPER SPACE

The ZOOM command is generally used to 'zoom in' on an area of a drawing to allow
the user to work on an enlarged area. The command is available in both Model and Paper
space, but the result is dependent on the mode being used.

Using the EX16_2 drawing with the additional viewport (Fig. 16.4):

1. Enter Model space and make the new viewport active.
2. Use the ZOOM command and zoom in on an area of the model using the window
   option – there is no apparent difference due to the size of the new viewport.
3. Undo the zoom effect.
4. Enter Paper space.
5. Use the zoom command to roughly window the complete new viewport.
6. The new viewport will be enlarged on the screen.
7. Enter Model space, and the enlarged viewport will be available to the user in Model
   space.

TILEMODE

1. With the modified EX16_2 drawing still on the screen, select from the menu bar
   File–New...
   prompt: save/discard/change – up to you!
   then: create new drawing dialogue box with acad name?
   prompt: **pick OK**
2. From the menu bar select **View**
                               Mview
                               4 Viewports
   prompt: ****Command not allowed unless TILEMODE is set to 0****
3. Select from the menu bar **View**
                             Tilemode
                             Off(0)
   prompt: New value for TILEMODE<1> : 0
             Regenerating drawing
   and: Paper Space icon.
4. Now select View–Mview–4 Viewports and create any four viewports of your choice
   on the screen.
5. Still in Paper space, select from the screen menu **SETTINGS**
                                                      next
                                                      VPORTS:
   prompt: ****Command not allowed unless TILEMODE is set to 1****
6. Enter Model Space, then select from the screen menu **MVIEW**
                                                        TILEMOD:
   prompt: New value for TILEMODE<0>
   enter: 1 <R>
7. The screen returns to a 'normal' display.
8. Now select SETTINGS–next–VPORTS–4.
9. A four viewport (model space) configuration is obtained.

Note

The TILEMODE variable determines which multi-screen drawing 'mode' can be used:
1. If Paper space is to be used, then TILEMODE is set to 0, and the MVIEW command is used to create untiled viewports.
2. If Model space is required, then TILEMODE is set to 1, and the VPORTS command is used to create tiled viewports.

SUMMARY

The following table compares the Model/Paper space drawing environments:

Model space	*Paper space*
Used to create model	Used to enhance drawing
Model can be modified	Model cannot be altered
Uses VPORTS command	Uses MVIEW command
TILEMODE set to 1	TILEMODE set to 0
Only current viewport plotted	All viewports plotted at once
Viewports are tiled	Viewports are untiled
Viewports cannot be modified	Viewports can be repositioned
WCS or UCS icon	Paper space icon
Zoom in active viewport	Zoom can 'fill screen'

MODEL VERSUS PAPER SPACE

Most new users to the Model/Paper space concept find it confusing as to what mode should be used to create multi-screen viewports. Both modes allow multiple viewports to be created and the user can set their own viewpoints with both modes.

Model space viewports are adequate for most applications, but cannot be 'modified' once created.

Paper space viewports allow the same flexibility as Model space viewports, but have advantages (as listed in the Summary table) which make them more suited to multiple viewport working.

My preference is to *create multi-viewports in paper space.*

ACTIVITY

1. Open drawing **ACT_8** from the 3DPACK directory.
2. A single viewport is displayed in the 'centre' of the screen – you may recognize it?
3. Using Paper space, create any viewport configuration of your own choice – Fig. 16.5 shows an example.

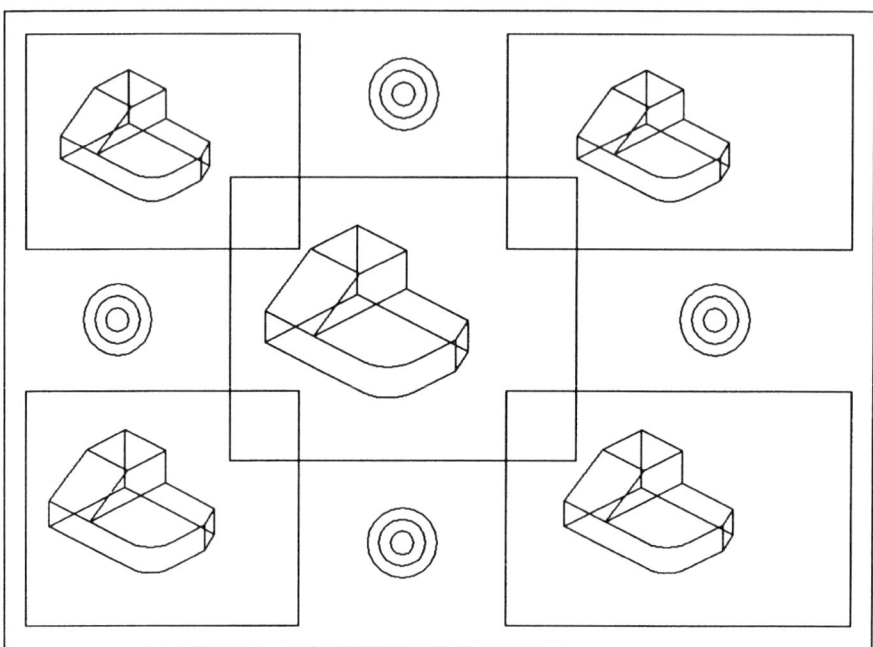

Fig. 16.5   ACT_8 with created viewports.

AN A3 MULTIVIEW STANDARD SHEET

The following sequence will create an A3 standard sheet (prototype drawing) for multiple viewport use. The procedure is the same for other sheet sizes (A2, A1, etc.). I have omitted certain settings such as units, text styles, dimvars, etc., as I would expect the user to have the ability to set these to their own requirements.

1. Begin a new drawing, accepting the ACAD prototype name.
2. Create the following layers:

BORDER	white	continuous
VP	yellow	continuous
MODEL	red	continuous

   Other layers can be added to suit, e.g. TEXT, DIMS, etc.
3. Use View–Tilemode to set Tilemode to Off (0).
4. Enter Paper space and make layer BORDER current.
5. Use the LINE command to draw from: 0,0

   to: @380,0

   to: @0,270

   to: @–380,0

   to: close
6. ZOOM All.
7. The resultant black border is the 'drawing sheet limits'.
8. Still in paper space, make layer VP current.
9. Using View–MView–4 Viewports:

   enter: 10,10 as the first point

   enter: 370,260 as the second point
10. The four required viewports are displayed in yellow inside the black border.
11. Enter Model space.
12. Set the following viewpoints:

top left	VPOINT 0,–1,0	FRONT view
top right	VPOINT 1,0,0	RIGHT side view
bottom left	VPOINT 0,0,1	TOP view
bottom right	VPOINT 'R' 315,30	3D view

13. These viewpoints are for first angle projection drawings. The user can set the viewpoint to suit third angle as required.
14. Make MODEL the current layer.
15. The standard sheet is now complete and can be used:

    (a) to create the model in Model space
    (b) to add extras (e.g. text) in Paper space.
16. Save this standard sheet as **3DSTD** in the 3DPACK directory – we may use it in a later exercise.

Note: we have still to 'centre' viewports, which will be discussed in the next chapter.

17

Centring viewports

One problem when working with multiple viewports is centring the model to 'line-up' with each other in the individual viewports. There are several ways to achieve this, and in this chapter we will investigate two of these methods:

(a) Zoom centre
(b) Zoom XP.

1. Open drawing **EX17_1** from the 3DPACK directory.
2. The screen displays a red wire-frame model in a four-viewport configuration, and it is obvious that the model is not at the viewport centre.

   Note: (a) the model was created using the 3DSTD standard sheet from the previous chapter.

   (b) The model is a cuboid block $200 \times 100 \times 110$ and has its origin at the point 0,0,0.

ZOOM CENTRE

This option requires the user to know:

(a) the coordinates of some suitable point on the model
(b) the model sizes, i.e. roughly the length, breadth and height.

As our model is $200 \times 100 \times 110$ and was created from the point 0,0,0 its 'centre point' has coordinates 100,50,55.

1. Ensure you are in model space.
2. Make the lower left viewport active.
3. Select from the screen menu **DISPLAY–ZOOM–Centre**

   prompt: Center point
   entert: **100,50,55 <R>**
   prompt: Magnification or height<297>
   entert: **300 <R>**

4. The top view is positioned at the centre of the viewport.

5. With the top left viewport active, enter **ZOOM** <R>
   prompt: All/Center...
   entert: **C** <R>
   prompt: Center point
   entert: **100,50,55** <R>
   prompt: Magnification or height<???>
   entert: **300** <R>
6. Repeat the ZOOM Centre selection in the other two viewports and
   enter: (a)   100,50,55 as the centre point
          (b)   300 as the magnification factor.
7. The model will be centred in each viewport as shown in Fig. 17.1.

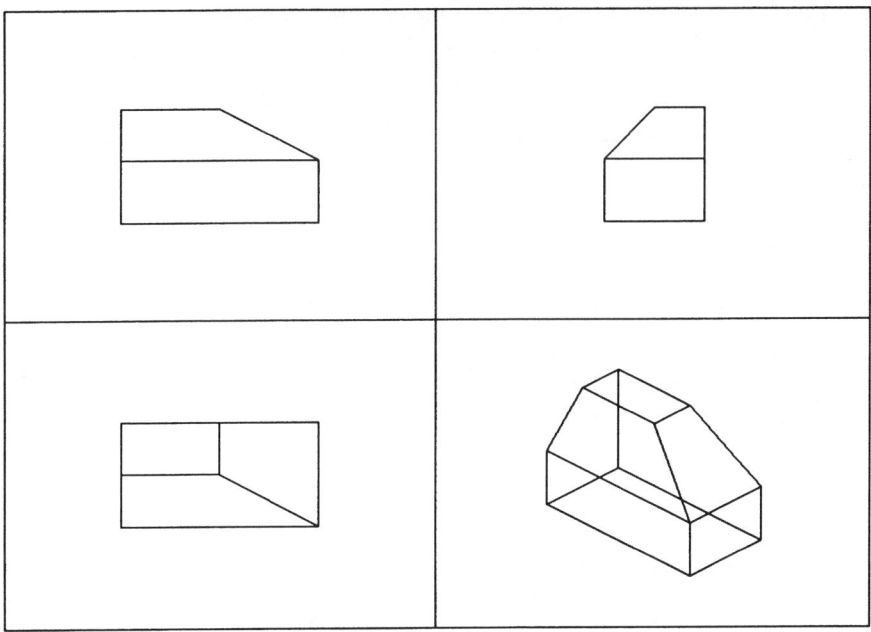

Fig. 17.1   EX17_1 with the ZOOM centre option at 300 maginification.

TASK

Using the same zoom centre point, investigate the following magnification factors:
   (a) 500,   (b) 150.

Note

1. The default magnification factor was <297> – step 3. If a value is entered:
   (a) greater than the default – display is reduced
   (b) less than the default – display is enlarged.
2. The same magnification factor should be used in the TOP, FRONT and SIDE viewports to 'line up' the model.
3. The magnification factor in the 3D viewport may be different from the other three viewports, due to the 3D model orientation.

ZOOM XP

This command produces the same effect as the Zoom–Centre option, and is considered easier to use. However, I find that the Zoom–Centre method gives the user 'greater control' over the display.

1. Open **EX17_1** again – the same drawing as before.
2. Make the lower left viewport active.
3. In each viewport ZOOM-Extents and the model will 'fill the screen'.
4. At the command line enter **ZOOM** <R>
   prompt: All/Centre/.../<Scale(X/XP)>
   enter: **0.5XP** <R>
5. Repeat the ZOOM-0.5XP sequence in the other three viewports.
6. The model is 'lined up' as shown in Fig. 17.2.

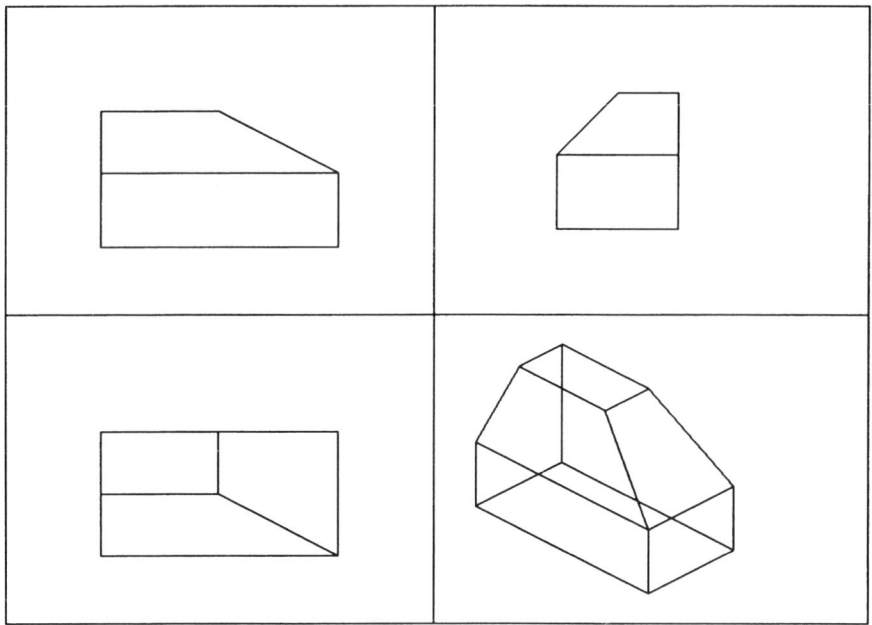

Fig. 17.2   EX17_1 with ZOOM 0.5XP.

TASK

Using the ZOOM command, investigate the following:
   (a) 0.75XP
   (b) 0.25XP
   (c) 1.5XP
Note: the larger the XP value, the larger the display.

SUMMARY

1. Model displays can be centred in the viewports using:
   (a) ZOOM–Centre
   (b) ZOOM XP
2. ZOOM–Centre requires a model centre point.
3. ZOOM XP is easier to use.

18

Three-dimensional polylines

In this chapter we will return to the 3D surfaces which we have been investigating earlier by discussing the 3DPOLY command.

A 3DPOLY is a polyline, but does not possess the drawing ability that a 2D polyline has, i.e. you cannot add curved segments to a 3D polyline. The 3D polyline can be edited in the same way as a 2D polyline.

EXAMPLE 1

1. Open drawing **EX18_1** from the 3DPACK directory.
2. The screen displays a three viewport configuration of two similar red wire-frame models with:
   (a) the left model – to demonstrate 3DFACES
   (b) the right model – to demonstrate 3DPOLY.
3. With the 3D viewport active, enter **HIDE** <R> and nothing happens.
4. Using the layer control dialogue box, thaw layer FACES and magenta faces are added to three of the surfaces of the left model.
5. Enter HIDE <R> and this model will be displayed with hidden line removal.
6. REGENALL to remove the HIDE effect.
7. Thaw layer POLY and two cyan shapes are added to two of the surfaces of the right model.
8. Enter HIDE again and:
   (a) the 3DFACE model has hidden line removal
   (b) the 3DPOLY model has no hidden line removal.
9. Remove the hide effect.

USING THE 3DPOLY COMMAND

1. Make layer POLY current and the 3D viewport active.
2. Select from the screen menu **SURFACES–3DPOLY**
   prompt: From point
   respond: **OSNAP END and pick pt 1**
   prompt: Close/Undo/<Endpoint of line>
   respond: **OSNAP END and pick pt 2**
   prompt: Close/Undo... and OSNAP END pt 3
   prompt: Close/Undo... and OSNAP END pt 4
   prompt: Close/Undo...
   enter: C <R> to end sequence.
3. The surface 1–2–3–4–1 will have a cyan 3DPOLY displayed.

USES FOR 3D POLYLINES

In their present context, 3D polylines can be used to add straight-edged shapes to surfaces for hatching (as can 3DFACES).

TASK

Using the three 3D polylines on the three surfaces, add hatching to the right model with the following hatch patterns:

(a) Stars   scale: 2   angle: 0
(b) Hexs   scale: 2   angle: 0
(c) Squares   scale: 2   angle: 0

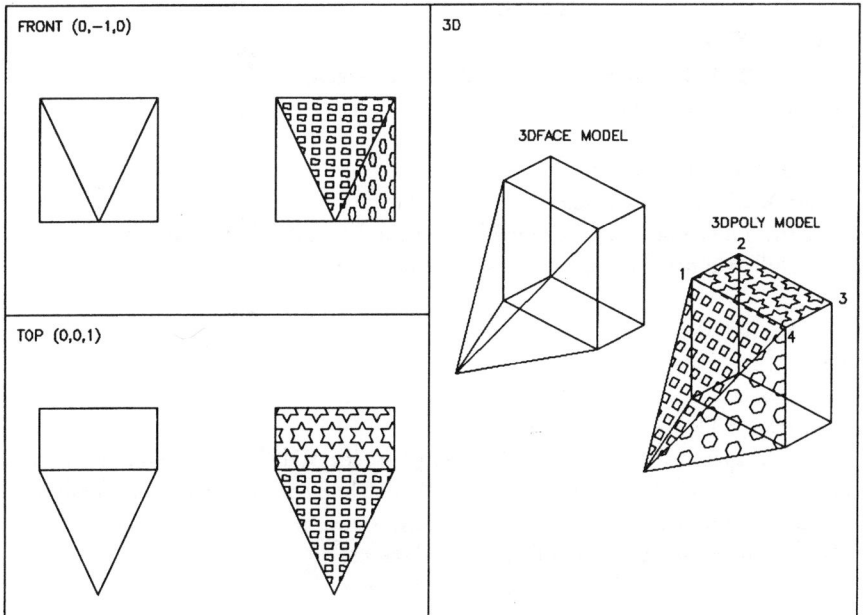

Fig. 18.1   EX18_1 after hatching the 3DPOLYs.

Notes

1. UCSs will be needed for this task, but you should be able to manage this on your own?
2. Use a HATCH layer
3. Result is as shown in Fig. 18.1.

EXAMPLE 2

In this example we will investigate another use for the 3DPOLY command which is the creation of a 'splined hill'.

1. Open drawing **EX18_2** from the 3DPACK directory.
2. The screen displays a four viewport configuration of the hill created at 50 interval heights using 3D polylines. Each height was created on its own coloured layer.
3. With the other three layers provided (levels 4,5,6), use the 3DPOLY command to enter the following x,y,z coordinate information:

Level 4			*Level 5*			*Level 6*		
195	115	200	195	135	250	195	145	300
270	145	200	250	155	250	225	160	300
255	185	200	230	175	250	195	160	300
175	185	200	180	170	250	close		
125	160	200	170	140	250			
160	120	200	close					
close								

4. From the menu bar select **Modify**
                         PolyEdit
   prompt: Select objects
   respond: **pick the red polyline at level 50** (the largest)
   prompt: polyedit options
   enter: **S** <R>
   prompt: polyedit options
   enter: **X** <R> to end sequence.
5. Use the Spline option of the PolyEdit command on the other 3D polylines.
6. The 'hill' will now be displayed as splined polylines.

TASK

1. Thaw layer TRAIL.
2. In model space (with layer ROUTE current) use the 3DPOLY command to plot the up and down routes for the hill. Use two 3D polylines, one for up and one for down.
3. Figure 18.2 displays the result of these added routes.
4. What happens to the routes of you 'spline' them?

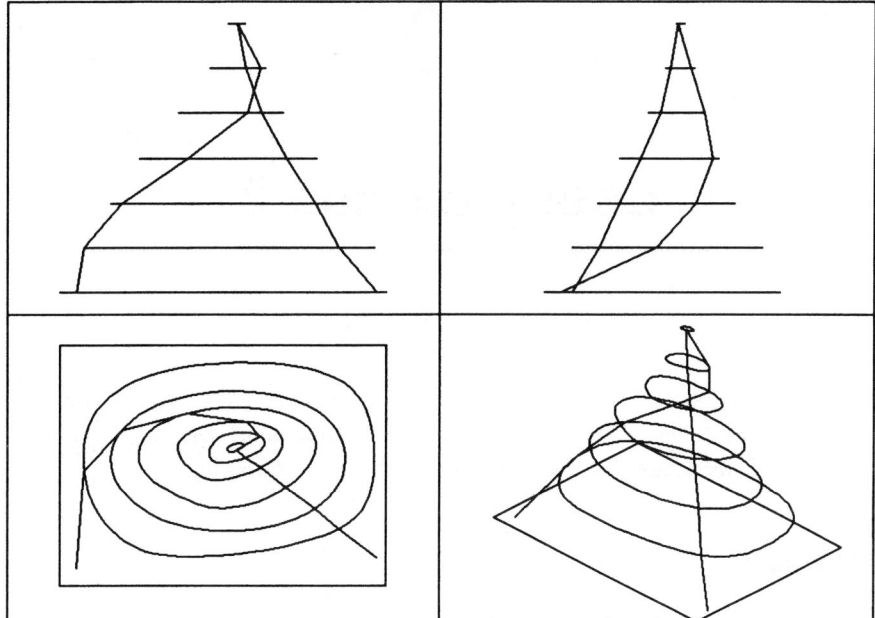

Fig. 18.2 The 3D splined hill (EX18_2) with the up and down routes.

SUMMARY

1. The 3DPOLY command can be used to:
   (a) outline straight surfaces for hatching
   (b) spline curve contours.
2. A 3DPOLY surface will not 'remove hidden detail' – 3DFACE will.
3. The 3DPOLY command does not permit arc segments to be added.

19

User exercise 2

This may be considered as a revision chapter, as it involves adding faces to a wire-frame model for hidden detail purposes. The exercise will also reinforce hatching in 3D and will be used to introduce the user to plotting multi-viewport drawings.

THE MODEL

1. Open drawing **USEX2** from the 3DPACK directory.
2. The screen displays a four-viewport configuration of a red wire-frame model – Fig. 19.1.
3. Investigate the layer control dialogue box, and note:
   (a) layers FACE1–FACE8 in different colours
   (b) layers HATCHSLOPE1 and HATCHSLOPE2.
4. Investigate the named UCSs.

TASK 1

1. Using the saved named UCS's with the eight FACE layers, use the 3DFACE command to face each surface of the model, i.e. one UCS per FACE layer.
   Note: (a) facing the L-shaped surface is interesting!
         (b) the model has eight 'surfaces':
   • one base
   • two rectangular vertical surfaces
   • two triangular/rectangular vertical surfaces
   • one L-shaped horizontal surface
   • two sloped surfaces.
2. Use the HIDE/SHADE commands in each viewport to give a quite pleasing coloured model.

TASK 2

Using the two hatch layers add hatching to the two sloped surfaces. *Note*: (a) you may have to 'freeze' some face layers for this(?), (b) I used U,45,5,N for the hatching.

TASK 3

1. Using the VPOINT Rotate option, change the viewpoint in the three 'non-3D' viewports, using positive angles, i.e. all four viewports should be 'looking down' on the model.
2. Save the four re-arranged viewports as **MVTASK** in the 3DPACK directory, and proceed to the next section on plotting multi-view drawings.

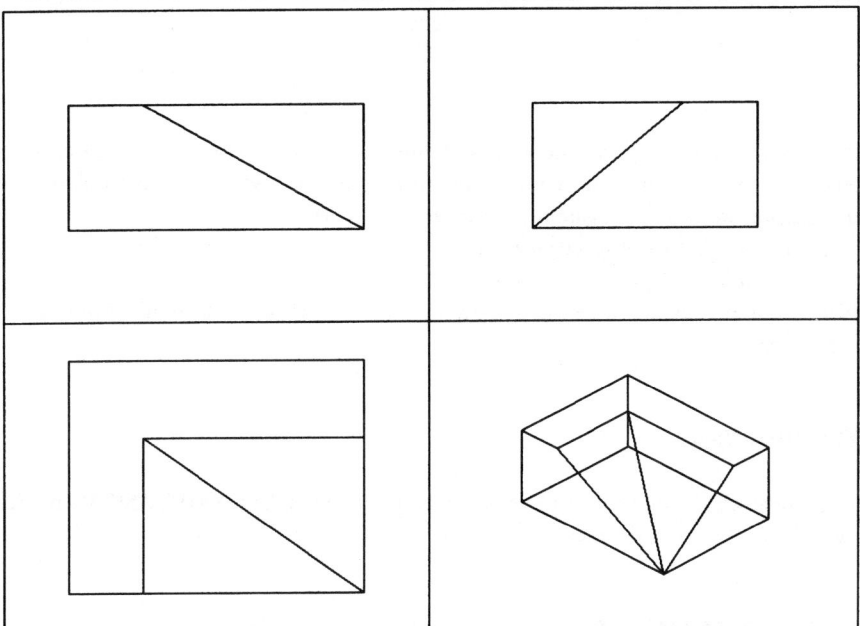

Fig. 19.1   USEX2 original model.

20

Plotting multiview drawings

Plotting drawings is a topic which I have found most AutoCAD users are frightened to tackle, and yet the final hard copy of the drawing is really what CAD is all about. In this chapter we will investigate how to plot 3D models:
- (a) as a single Model space viewport
- (b) as multiple Paper space viewports.

Note: if you do not have access to a plotter, then it may be advisable to miss this chapter completely.

THE DRAWING

The model which will be used to demonstrate plotting will be the MVTASK 3D model (with hatching) from the previous chapter.

THE PLOT COMMAND

The command to plot a drawing can be activated:
- (a) from the root screen menu with **PLOT**:
- (b) from the menu bar with **File–Plot...**
- (c) from the keyboard by entering **PLOT** <R>

Each method results in the Plot Configuration dialogue box, which can be rather off-putting to new users. The dialogue box gives details of:
1. The plotter device.
2. The paper size and orientation.
3. The pen parameters.
4. The plot scale, rotation and origin.
5. Plot type – display, extents, limits and hide.
6. Plot preview.

Plotting is one of the most individualistic functions within AutoCAD, i.e. some users plot in inches to Display on A2-paper full size, while other users may plot Extents in millimetres to fit A1 paper. I'm sure that you can appreciate that it is impossible to cover all possible variations, but I hope that you will obtain enough information to enable you to plot your 3D models to your requirements.

The plotter device

This is what will do the plotting. Ideally it will always be set to the plotter which is used, but some users may have access to more than one plotter, e.g. pen, laser, thermal, etc. By selecting the Device and Default selection box, the user can then pick the plotter to be used from those listed. The plotter device drivers have to be 'loaded' in the AutoCAD system.

The paper size and orientation

By selecting the Size... box, the paper size can be chosen. As all our work is on A3 paper in millimetres, then A3, MM should be displayed here.

Note: A4 laser printers will not accept A3 paper (obviously) and thus A4 paper size will be required.

The pen parameters

This allows pens to be positioned in different 'slots' if you have a multi-pen plotter. I generally leave the settings as they are, and if coloured plots are required, I use to AutoCAD coloured numbers, i.e. red 1, yellow 2, green 3, etc.

The plot scale, rotation and origin

(a) The plot origin is normally 0,0 as this is where most drawings are plotted from.
(b) The rotation is normally 0 or 90 and is dependent on how the paper is loaded into the plotter, i.e. long side or short side.
(c) Plot scale: the most confusing option. We generally want full-sized plots, so the scale is 1=1, but this obviously depends on the actual drawing and the paper size being used.
(d) Scaled to fit: when selected (X in box) the plot will be scaled to fit the paper size being used. Useful with 3D models.

Plot type – display, extents, limits and hide

Display, Extents, Limits can give different plots. I generally plot to Display, i.e. 'what you see is what you get'. It is advisable to try the various options for yourself.

Hide Lines: when selected (X in box) hidden detail will be removed from Model space viewport models, e.g. extrusions, faced, surfaced, revolved.

Plot preview

The full preview option is very useful as it will display the drawing relative to the paper size. It should be used at all times prior to obtaining the plot, as it will allow corrections in the plot scale, or plot type to be made.

The above is a rough idea of the PLOT command.

MODEL SPACE PLOT

In Model space *only the active viewport can be plotted*, i.e. it is only possible to plot a single model space viewport on a sheet of paper.

1. Ensure the MVTASK drawing is displayed on the screen.
2. Make the lower right viewport active.
3. Activate the PLOT command and set your plot variables, e.g.
   (a) device : ???????
   (b) size: A3 paper in MM
   (c) origin: 0,0 with plot rotation : 0
   (d) plot Display with Hide lines (X in box)
   (e) plot scaled to Fit, i.e. X in box
   (f) preview: full
4. When all variables have been set for your system, pick OK.
5. Paper to be positioned in plotter, then <RETURN>.
6. The active viewport will be plotted with hidden line removal as shown in Fig. 20.1.
7. Plot the other viewports.

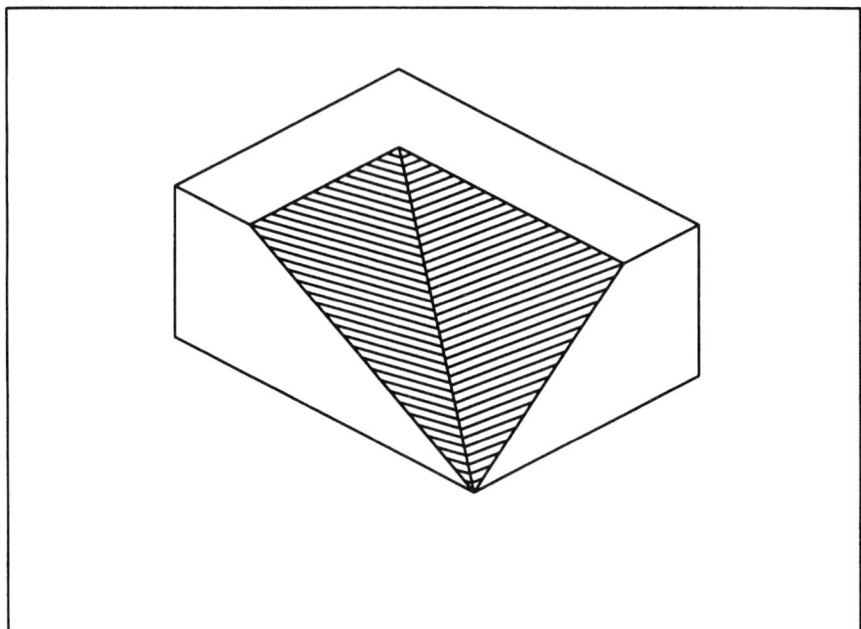

Fig. 20.1 MVTASK drawing plotted as a single model space viewport using HIDE option from the Plot Configuration dialogue box.

PAPER SPACE PLOT

In Paper space it is possible to *plot any viewport configuration* on a single sheet of paper.
1. Ensure MVTASK drawing still displayed.
2. Enter Paper space with PS <R>.
3. From the menu bar select **View**

> **Mview**
> **Hideplot**
> **ON** – from screen menu

    prompt: Select objects

    respond: **pick the 4 yellow borders of the individual viewports until each outline is highlighted then <R>**

    prompt: Command line returns

4. Now use the PLOT command as normal. The Hide Lines option can be 'turned off', i.e. no X.
5. The plot of the multi-viewports will be obtained as shown in Fig. 20.2.

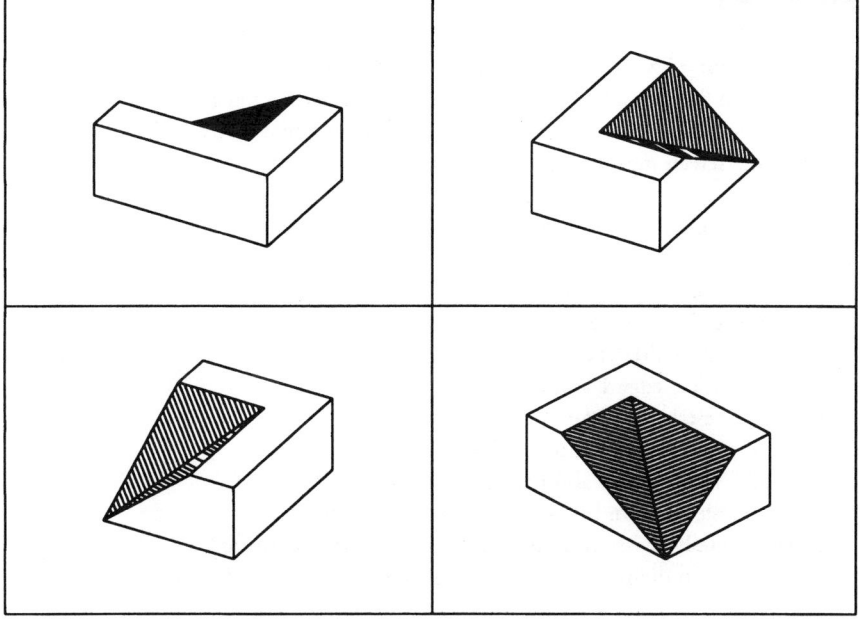

Fig. 20.2 MVTASK plotted in Paper space using HIDEPLOT ON.

SUMMARY

1. In Model space, only the active viewport can be plotted at any one time. Hidden lines are removed from the plot by using the Hide Lines option box.
2. Paper space allows all viewports to be plotted at once. Hidden lines are removed from a multi-view plot by using the sequence MVIEW–HIDEPLOT–ON and selecting individual viewport borders.

21

Edge surfaces

The edge surface command (**EDGSURF**) produces a very similar effect to a 3D mesh, i.e. a net effect. The main difference is that the user does not have the tedious task of entering individual vertex coordinates.

EXAMPLE 1

1. Open drawing **EX21_1** from the 3DPACK directory.
2. The screen displays a four viewport configuration with a red square.
3. Ensure layer EDGE (cyan) is current and the lower right viewport is active.
4. From the screen menu select **SURFACES**
                    Surftb1:
   prompt: New value for SURFTAB1 <18>
   enter: **12 <R>**
5. Repeat for Surftb2 and set to 10.
6. Now select **SURFACES**
                    EDGSURF:
   prompt: Select edge 1 and pick line 1–2
   prompt: Select edge 2 and pick line 2–3
   prompt: Select edge 3 and pick line 3–4
   prompt: Select edge 4 and pick line 4–1
7. A cyan mesh is drawn between the four selected lines.
   Note: (a) mesh divisions in the *Y*-axis direction are 12–surftb1 and
           (b) mesh divisions in the *X*-axis direction are 10–surftb2.

EDITING AN EDGE SURFACE

An edge surface mesh is edited in a similar way to a 3D mesh, i.e. vertices can be moved and the mesh surface can be smoothed.

1. Make the lower left viewport active.
2. Thaw layer CHANGEVERTICES which will display several donut shapes with letters beside them.
3. Enter Paper space and zoom in on the lower left viewport, then return to Model space with the lower left viewport active.

4. From the menu bar select **Modify**

    PolyEdit

    prompt: Select object

    respond: **pick the cyan mesh**

    prompt: Edit Vrtx/…

    enter: **E <R>** – for the edit vertex option

    prompt: Vertex(0,0), Next/…

    and: white cross at bottom left vertex?

    enter: **N <R>** repetitively until the white cross is at the vertex denoted with the

              donut lettered a.

              (*Note*: try the U,D,R,L entry options)

    then: **M <R>** – when X at letter a

    prompt: enter new location

    enter: **@0,0,–30 <R>**

    prompt: Vertex(?,?). Next/…

5. Using the N,U,D,L,R move options, move the white cross to the other lettered donuts, then use the **M** option and enter the following new locations for the letter stated:

    b: @0,0,40   c: @0,0,–50

    d: @0,0,20   e: @0,0,30

6. Continue moving the cross to the other lettered donuts and enter the following new locations:

f: @0,0,–5	g: @0,0,–15	h: @0,0,–20
i: @0,0,–5	j: @0,0,–10	k: @0,0,–25
l: @0,0,–30	m: @0,0,–40	n: @0,0,–15
o: @0,0,–30	p: @0,0,–50	q: @0,0,–40
r: @0,0,–20	s: @0,0,–50	t: @0,0,–60
u: @0,0,–70	v: @0,0,–90	

7. Enter Paper space and zoom Previous, and return to Model space. The edge surface will be displayed with the edited vertices in all viewports.

8. Freeze layer CHANGEVERTICES.

9. At the command line enter **PEDIT <R>**

    prompt: Select polyline

    respond: **pick any point on the mesh**

    prompt: Edit Vrtx/…

    enter: **S <R>**

    then: **X <R>** to end sequence.

10. The edge surface will be 'smoothed' about the altered vertices as shown in Fig. 21.1.

11. Try HIDE, SHADE in each viewport.

12. Multi-view plot?

13. Save if required.

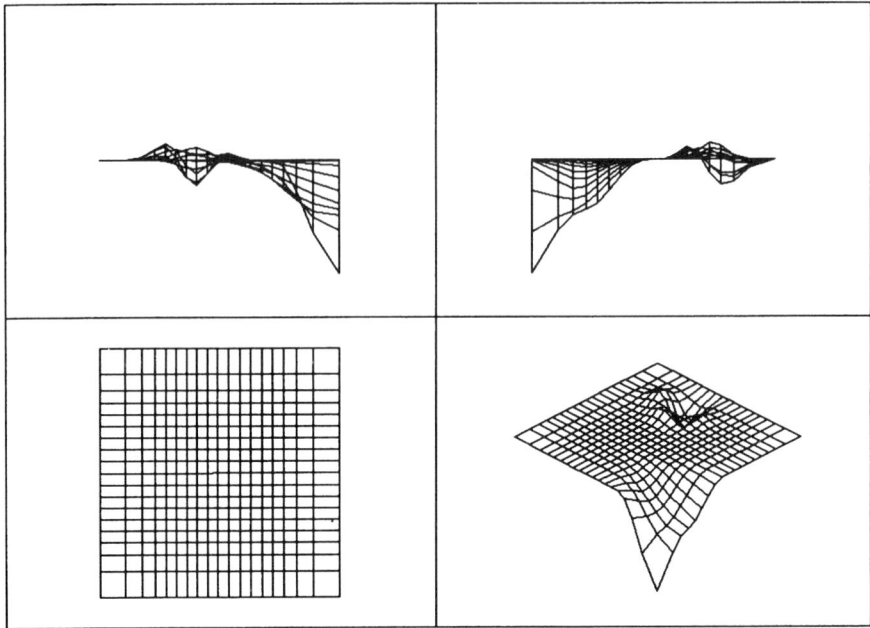

Fig. 21.1   EDGE SURFACE example 1 after Edit Vertex and Smooth operations.

EXAMPLE 2

1. Open drawing **EX21_2** from the 3DPACK directory.
2. Screen displays a green 'circular' edge surface mesh. Any idea how the circular edge surface could have been created? Think arcs!

TASK

Can you edit the vertices to produce an effect similar to that displayed in Fig. 21.2, i.e. a raised centre with a 'trough' around it?

> *Hint*: thaw layer ASSIST and use the coloured donuts with the following changes:

    red: @0,0,50
    yellow: @0,0,20
    blue: @0,0,−20
    cyan: @0,0,−30

EXAMPLE 3

1. Open drawing **EX21_3** from the 3DPACK directory.
2. Screen displays four coloured 3D polylines drawn in a rather strange pattern in a centred four viewport configuration. Paper space is active?
3. Change to Model space.

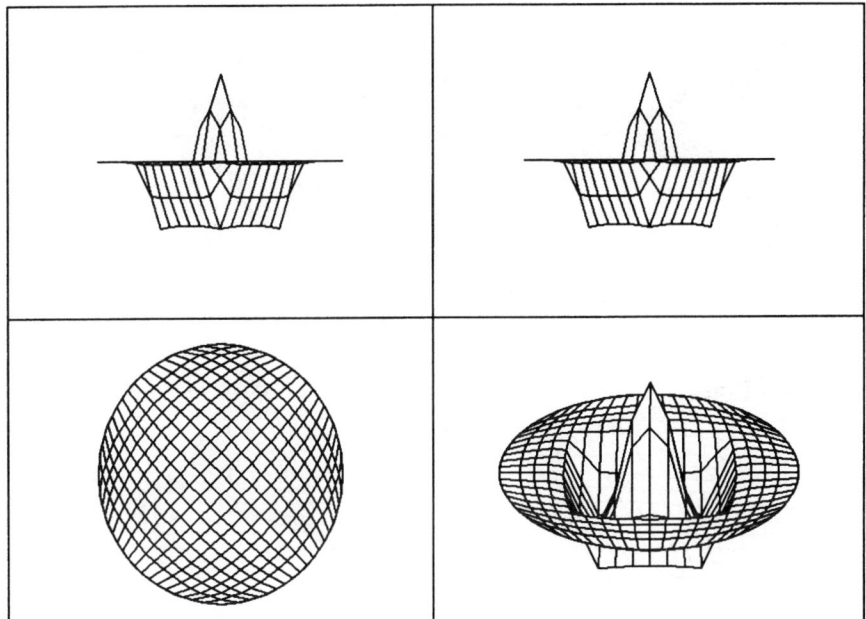

Fig. 21.2 EDGE SURFACE example 2 after Edit vertices and Smooth operations.

TASK

1. Edge surface the 3D polylines, picking in the order red, green, blue and yellow.
2. Edit to give a smooth surface as shown in Fig. 21.3.
3. Change the four viewpoints to different values similar to Fig. 21.4.

Hint: layer DONUT has four coloured (2D) donuts drawn at the four mesh vertices. This may help to give you some idea of the orientation of the edge surface then the viewpoints have been altered.

SUMMARY

1. To use the edge surface command, four **TOUCHING** entities are needed.
2. The entities can be lines, arcs, polylines or 3D polylines.
3. The actual edge surface mesh is controlled by the SURFTB (surface tabulation) variable, and:
   (a) surftb1 is the *Y*-direction control
   (b) surftb2 is the *X*-direction control.
4. Edge surfaces can be used with HIDE and SHADE.

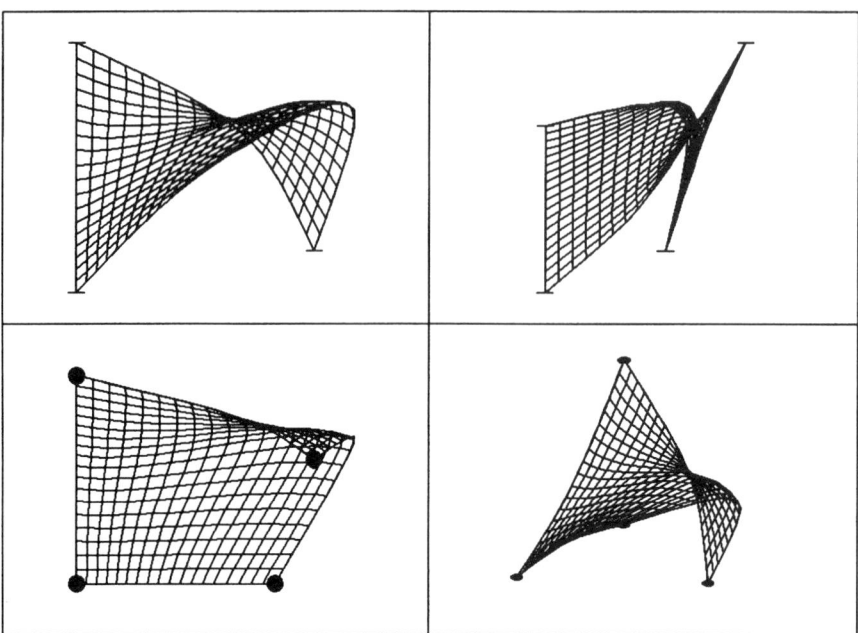

Fig. 21.3   EDGE SURFACE example 3.

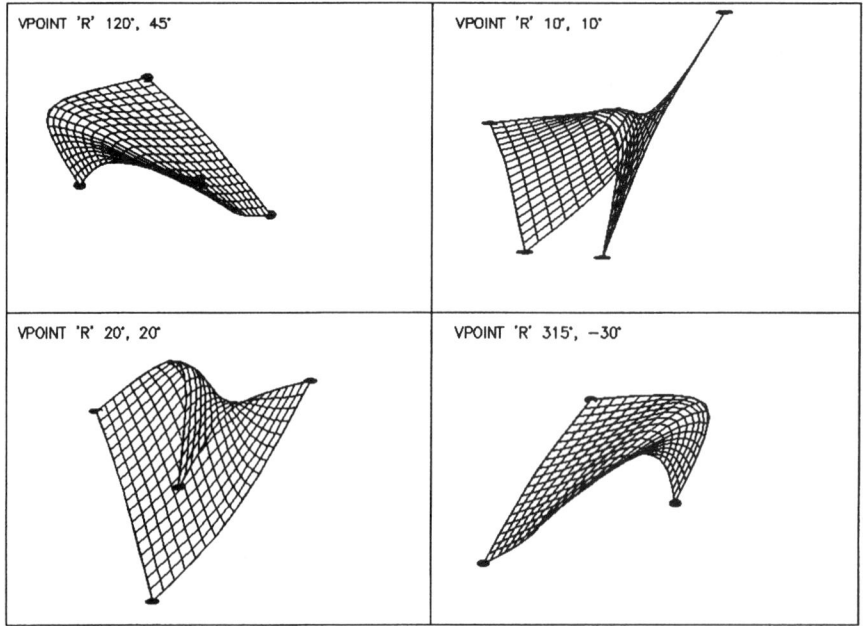

Fig. 21.4   EDGE SURFACE example 3 at different VPOINT 'R' angles.

22

Tabulated surfaces

Tabulated surfaces add a 'direction vector' to a selected 'path' and can be used in 2D or 3D. They are useful in making a 3D wire-frame model into a surface model.

When the command is activated, the user has two prompts:
1. The path curve
2. The direction vector.

EXAMPLE 1

1. Open drawing **EX22_1** from the 3DPACK directory.
2. The screen will display several red entities with some black donuts and text. The donuts and text are for reference only.
3. Ensure TABSURF1 is the current layer.
4. Refer to Fig. 22.1 for the referenced points which follow.
5. From the menu bar select **Draw**
                    3D Surfaces
                    Tabulated Surface
   prompt: Select path curve
   respond: **pick a pt 1 on the line indicated in figure (a)**
   prompt: Select direction vector
   respond: **pick a pt 2 on the line indicated in (a)**
6. Repeat the TABSURF command, selecting the path curve (point 1) and direction vectors (point 2) indicated in Fig. 22.1.
7. The figures in Fig. 22.1 have been designed as follows:
   (a)–(d): line as path curve, line as direction vector
   (e)–(g): arc as path curve, line as direction vector
   (h)–(j): circle as path curve, line as direction vector.
8. Finally alter the value of SURFTB1 to 4, and TABSURF the lines of (k).

Notes

1. The path curve can be a line, arc or circle.
2. The direction vector *can only be a line.*
3. The 'direction' of the tabulated surface depends on the end of the direction vector line selected.
4. The tabulated surface 'segments' is controlled by SURFTB1.

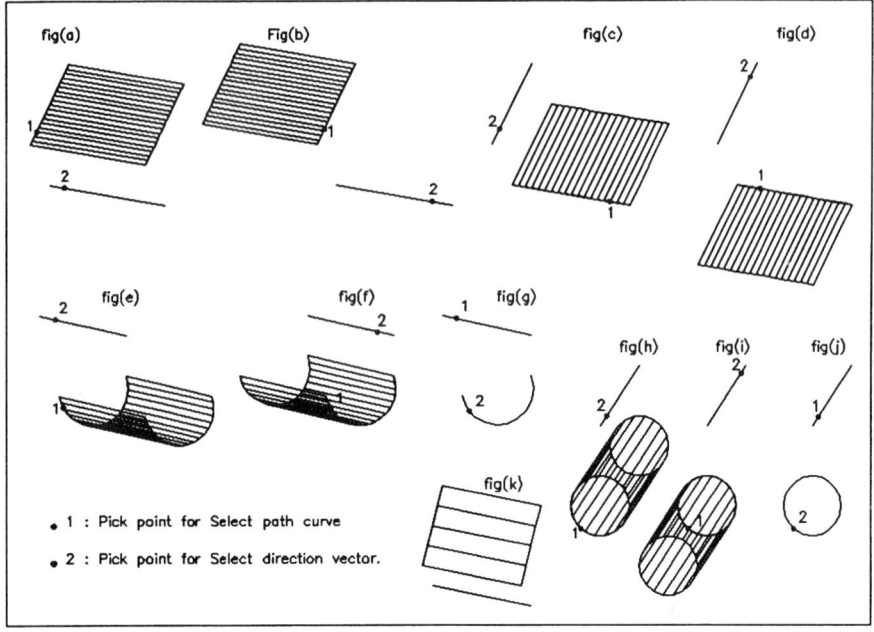

Fig. 22.1   Tabulated surface EX21_1 with lines, arcs and circles in 2D.

EXAMPLE 2

1. Open drawing **EX22_2** from the 3DPACK directory.
2. The screen displays shapes drawn:
   (a) as a closed polyline
   (b) as a polyline-arc segment.
3. Ensure layer TABSURF is current.
4. From the screen menu select **SURFACES**
                              TABSURF:
   (a) pick the polyline shape as the path curve (pt 1)
   (b) pick the line at the end points indicated as the direction vector (pt 2).
5. Repeat until the four polyshapes have been tabulated – Fig. 22.2.
6. Try and pick the line as the path curve, and one of the polyshapes as the direction vector.

EXAMPLE 3

1. Open drawing **EX22_3** from the 3DPACK directory and refer to Fig. 22.3.
2. The screen displays a red wire-frame model created from lines and arcs.
3. Ensure TABSURF layer is current.

Example 3   109

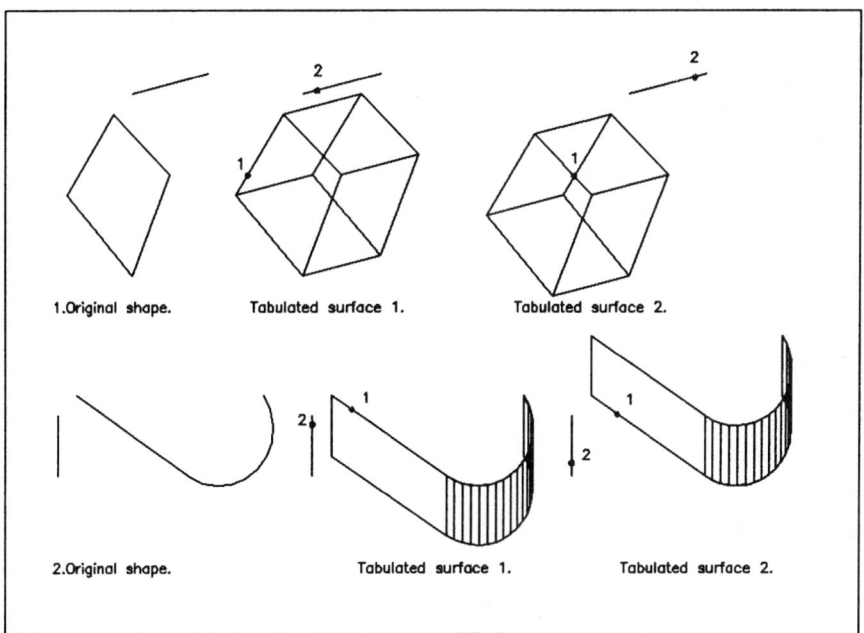

Fig. 22.2   Tabulated surface example 2 – 2D polylines.

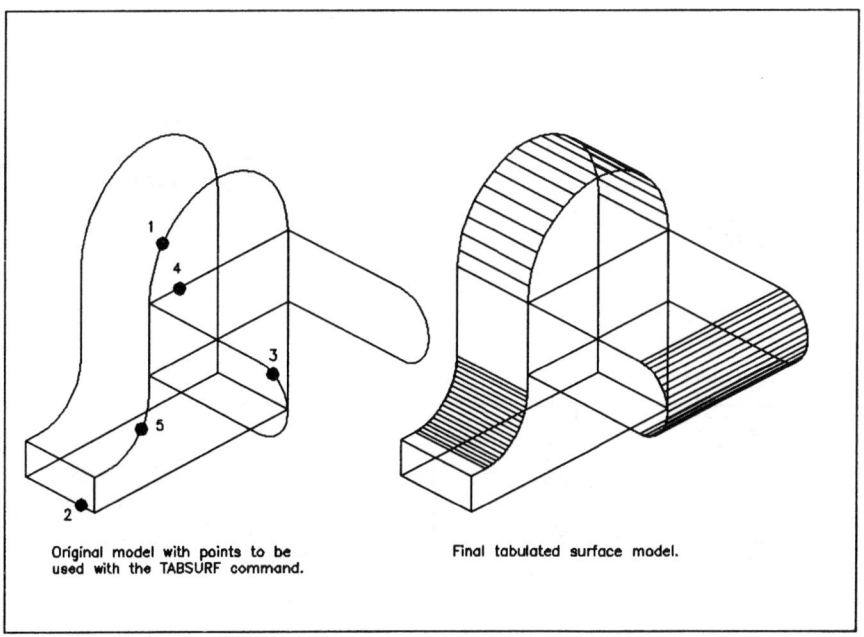

Fig. 22.3   Tabulated surface example 3 – wire-frame model.

4. Use the TABSURF command three times to add tabulated surfaces to the model with the following:
   (a) pt 1: path curve
       pt 2: direction vector
   (b) pt 3: path curve
       pt 4: direction vector
   (c) pt 5: path curve
       pt 2: direction vector.
5. Try HIDE and SHADE.

TASK

Can you set a centred four-viewport configuration for the tabulated surface 3D model similar to Fig. 22.4?

SUMMARY

1. Tabulated surfaces require a path curve and a direction vector.
2. The path curve can be a line, arc, circle, polyline.
3. The direction vector must be a line.
4. Tabulated surfaces can be used in 2D or 3D.
5. The tabulated surface direction depends on the direction vector 'pick point'
6. HIDE and SHADE are possible with tabulated surfaces.

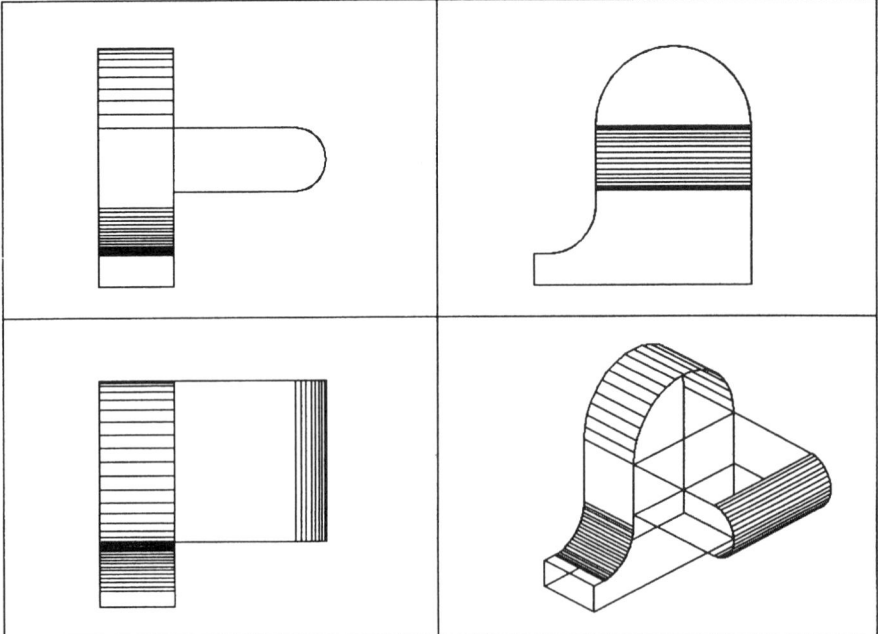

Fig. 22.4   Four viewports of tabulated surface example 3.

ACTIVITY

It has been some time since you have had an activity.

1. Open the activity **ACT_9** from the 3DPACK directory.
2. Use TABSURF to add tabulated surfaces to the 3D wire-frame model which is displayed. The command must be used eight times (?) and no help is given about what should be selected for the path curve or the direction vertex, but you should not have any trouble with the exercise.
3. Change the viewpoints similar to Fig. 22.5.

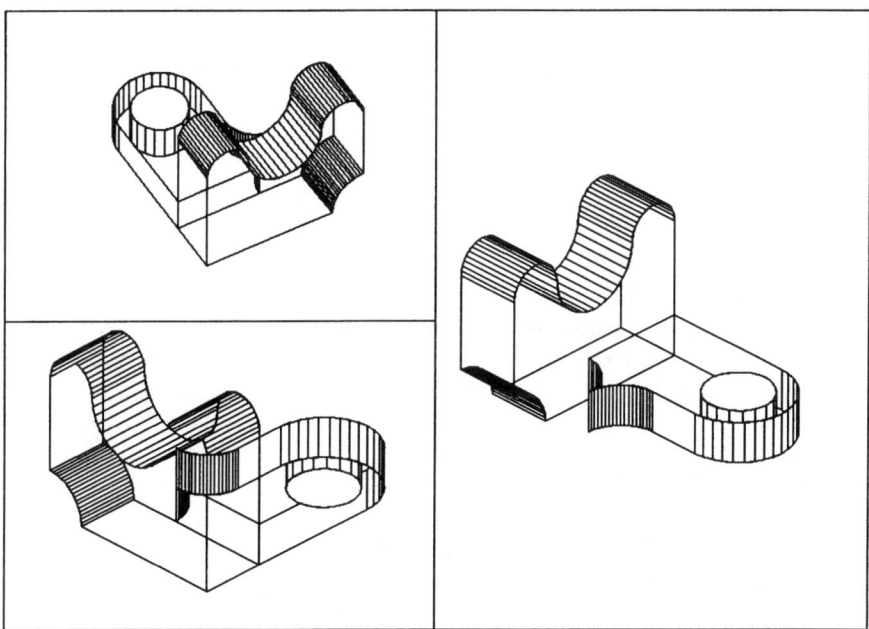

Fig. 22.5   Activity 9 with tabulated surfaces at three different viewpoints.

23

Ruled surfaces

Ruled surfaces allow wire-frame models to be 'converted' into surface models by adding surfaces between two selected entities. These entities can be:
(a) two lines
(b) two arcs
(c) an arc and a line
(d) two circles
(e) two polyline segments
(f) a circle and a point.

The command *cannot* be used between a line and a circle, or an arc and a circle.

The final effect of the ruled surface depends (as with tabulated surfaces) on where the points are selected on the desired entities.

EXAMPLE 1 – 2D LINES, ARCS AND CIRCLES

1. Open drawing **EX23_1** from the 3DPACK directory.
2. The screen displays a single viewport with several red entities. Black numbers are displayed for reference only. Layer RULSURF (blue) should be active.
3. Refer to Fig. 23.1
4. From the screen menu select **SURFACES**
                                RULSURF:
   prompt: Select first defining curve
   respond: **pick the end of line indicated by pt 1 in figure (a)**
   prompt: Select second defining curve
   respond: **pick the end of line indicated by pt 2 in (a)**
5. A blue surface will be drawn between the two selected entities.
6. From the menu bar select **Draw**
                                3D Surfaces
                                Ruled Surface
   prompt: Select first defining curve
   respond: **pick end pt 1 in (b)**
   prompt: Select second defining curve
   respond: **pick end pt 2 in (b)**

Example 1   113

7. At the command line enter **RULESURF <R>**
   prompt: Select first defining curve
   respond: **pick pt 1 of arc indicated in (c)**
   prompt: Second defining curve
   respond: **pick pt 2 of arc indicated in (c)**
8. Using the screen menu, menu bar or keyboard entry method, use the ruled surface command selecting the first and second defining curves as indicated in Fig. 23.1:
   (d) two arcs
   (e) line and arc
   (f) two circles
   (g) point and circle
   (h) circle and line.
9. Ruled surfaces will be added between the selected entities, with the exception of (h). With the line and circle, an error message is displayed in the command prompt area: **'cannot mix closed and open paths'.**
10. Finally, select **Surftb1** from the screen menu and:
    prompt: New value for SURFTAB1<18>
    enter: 6 <R>
11. Now rule surface the two lines in (i).
12. Your screen should resemble Fig. 23.1.

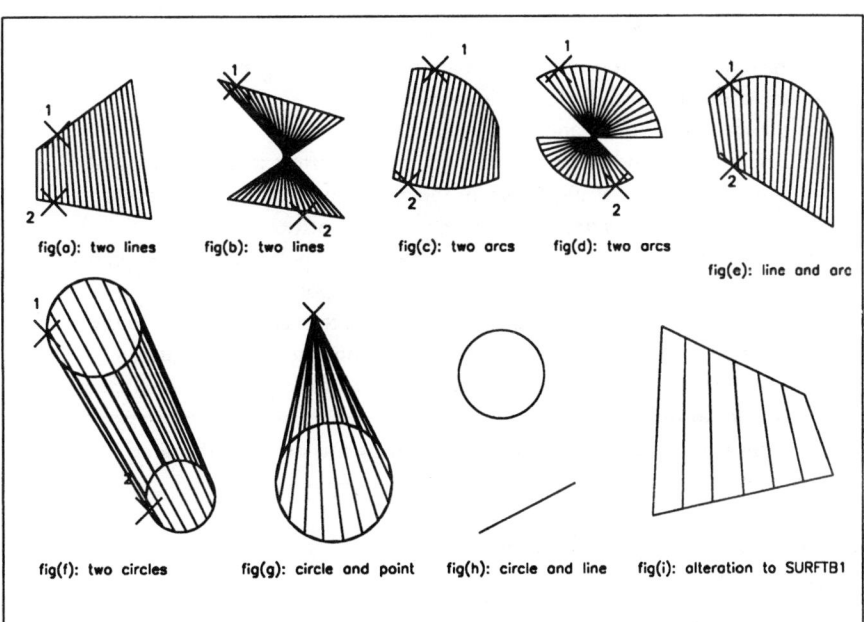

fig(a): two lines   fig(b): two lines   fig(c): two arcs   fig(d): two arcs

fig(e): line and arc

fig(f): two circles   fig(g): circle and point   fig(h): circle and line   fig(i): alteration to SURFTB1

Fig. 23.1   Ruled surface example 1.

CLOSED AND OPEN PATHS

1. The RULESURF command can only be used between:
   (a) two open path entities, e.g. lines, arcs
   (b) two closed path entities e.g. circles, points.
2. The command cannot be used between open and closed path entities, e.g. between a line (open) and a circle (closed).

EXAMPLE 2 – 2D POLYLINES

1. Open drawing **EX23_2** from the 3DPACK directory.
2. The screen displays several red polyline entities with black donuts (for reference only).
3. Refer to Fig. 23.2.
4. Activate the ruled surface command and pick the two defining curves at the end points indicated by the donuts in (a)–(d). The polylines are surfaced.
5. Repeat the ruled surface command for (e) and (f), selecting the polyline square as the first defining curve and the circle as the second defining curve.
   (e) open/closed message
   (f) ruled surface added.
6. (a) The polyline square in (e) was drawn as four lines, and the sequence was ended with <RETURN>.
   (b) The polyline square in (f) was drawn as three lines, and the square completed with the CLOSE option, i.e. **C <R>**. A closed polyline sequence is considered as a closed path, and therefore (f) will rule surface. A polyline which is not closed is an open path, thus (e) will not rule surface.
   This concept is important to remember when using polylines with the ruled surface command and circles are involved.
7. Rule surface (g) and (h) and:
   (g) the polygon and circle are surfaced, but the effect is not as expected. This is due to the circle being drawn in the opposite 'sense' from the polygon
   (h) ruled surface as expected – two closed path entities.
8. Rule surface (i) and (j) which consist of a polyline square drawn from 1–2–3–4–close, and a circle drawn as two closed polyline arcs
   (i) the entities are surfaced, but the effect is not correct. This is due to the polyline arcs being drawn in the opposite 'sense' from the polyline square
   (j) effect as expected, as the polyline square and the polyline arc circle are both drawn in the same 'sense'.
9. Now rule surface (k), which is two polyline segments drawn in the same 'sense'. This type of selection is useful in 3D.
10. Your screen should now resemble Fig. 23.2.
11. Polylines can be rule surfaced, but:
    (a) must be closed
    (b) must be drawn in the same direction
    (c) start points are important.

Example 3   115

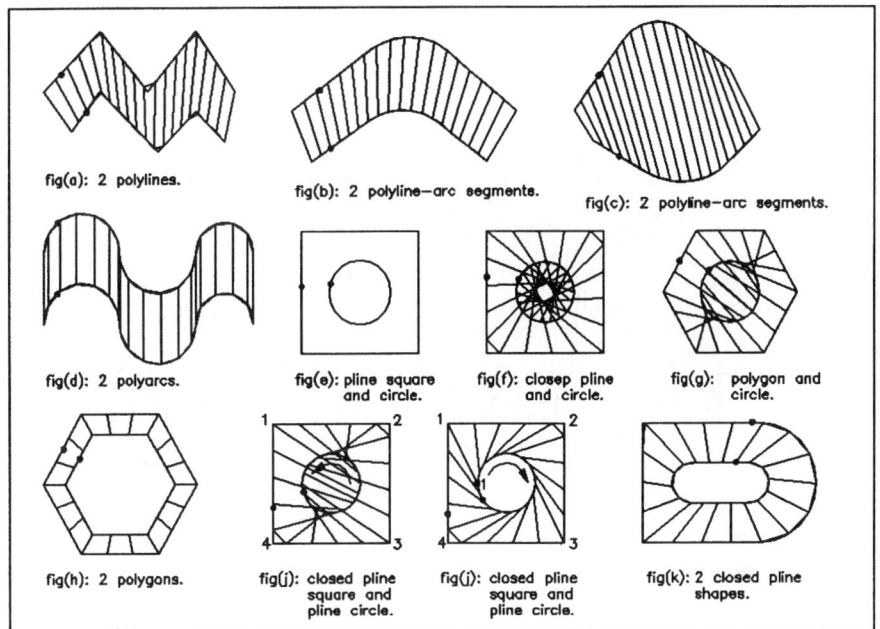

fig(a): 2 polylines.

fig(b): 2 polyline–arc segments.

fig(c): 2 polyline–arc segments.

fig(d): 2 polyarcs.

fig(e): pline square and circle.

fig(f): closep pline and circle.

fig(g): polygon and circle.

fig(h): 2 polygons.

fig(j): closed pline square and pline circle.

fig(j): closed pline square and pline circle.

fig(k): 2 closed pline shapes.

Fig. 23.2   Ruled surface example 2 – polyline.

EXAMPLE 3 – RULE SURFACES IN 3D

The ruled surface command in 3D is very powerful and can be used with lines, arcs or polyline shapes. When used, the command will convert a wire-frame model into a surface-model. The HIDE and SHADE commands are then useful.

1. Open drawing **EX23_3** from the 3DPACK directory.
2. The screen displays two red wire-frame models:
   (a) model 1 – drawn from lines and arcs
   (b) model 2 – drawn from polyline line/arc segments.
3. Thaw layers SURFACE1, SURFACE2 and SURFACE3 which will display three coloured ruled surfaces on the wire-frame 'arches'.
4. Thaw layers SURFACE4-6, which will display coloured surfaces on three of the model 'flats'. *Question*: the ruled surfaces on the arches should be obvious. How were the ruled surfaces created on the 'flats', especially the blue surface?
5. Try HIDE and SHADE – effect of the ruled surface should be obvious.
6. Thaw layers POLYSURF1, POLYSURF2 and POLYSURF3 to add three ruled surfaces to Model 2. HIDE and SHADE – interesting?
7. Figure 23.3 displays the two surfaced models with HIDE.

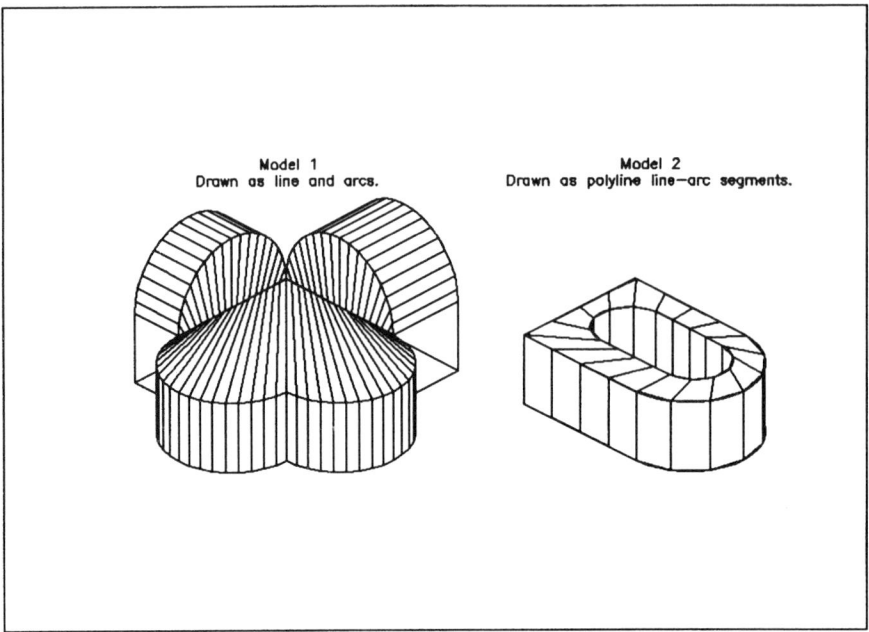

Fig. 23.3   Ruled surface example 3 – 3D wire-frame models.

SUMMARY

1. Ruled surfaces can be added between lines, arcs and polylines.
2. The command can be activated from the screen menu, the menu bar or by keyboard input.
3. Two 'defining curves' are required to be selected.
4. The curves must both be open or closed.
5. Open and closed paths cannot be used with the command.
6. Polyline line/arc segments are very useful.
7. When a large 3D wire-frame model is to be surfaced, different layers should be used for each surface, e.g. top, side, front, etc. When a layer has been used for surfacing, it should be frozen before another layer is used.
8. The number of ruled surface 'segments' is controlled by the SURFTB1 variable.

ACTIVITY

This is an interesting activity, and typical of the use which can be made with the ruled surface command.

1. Open drawing **ACT_10** from the 3DPACK directory.
2. The screen displays a four viewport (all the same) configuration on a red wire-frame model drawn with polyline line and arc segments and lines.
3. Rule surface the model using the four layers provided:
   S1: for the top surfaces
   S2: for the right side surfaces
   S3: for the left side surfaces
   S4: for the bottom surfaces
   Hint:　1. make layer S1 current and rule surface the 'tops'
   　　　　2. make layer S2 current
   　　　　3. freeze layer S1
   　　　　4. rule surface on the right surfaces
   　　　　5. continue with the other layers/surfaces.
4. Set the viewports to show a top, front and side view.
5. Centre each viewport – consider the model as a cuboid with sizes 350 × 350 × 575.
6. HIDE and SHADE each viewport – colour effect as expected?
7. Figure 23.4 displays the surfaced model with HIDEPLOT ON.

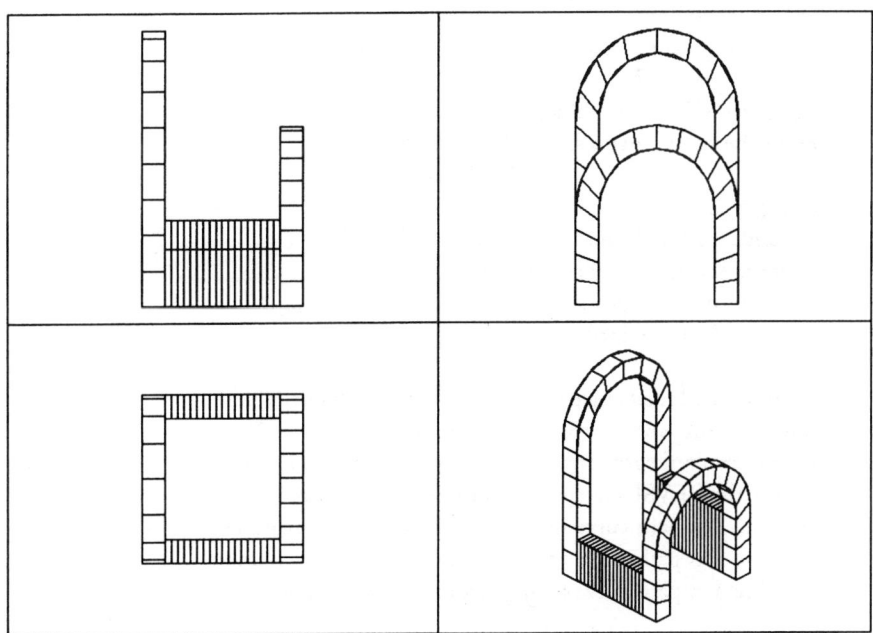

Fig. 23.4 Wire-frame model for RULED SURFACE activity as four viewports (with HIDEPLOT ON).

24

Revolved surfaces

This command allows the user to produce complex 3D surface models with two selections:

(a) a path curve

(b) an axis of revolution.

Once the two selections have been made, the user then enters a start angle and an included revolution angle.

EXAMPLE 1 – LINES AND POLYLINES

1. Open drawing **EX24_1** from the 3DPACK directory.
2. The screen displays a two viewport configuration with:
   (a) four red lines
   (b) a red polyline shape
   (c) two green lines
   (d) layer SURFACE (blue) current.
3. From the screen menu select **SURFACES**

                           REVSURF:

   prompt: Select path curve

   respond: **pick red line 1** – note *no* <RETURN>!

   prompt: Select axis of revolution

   respond: **pick the green line** (at 'bottom' end)

   prompt: Start angle<0>

   enter: **0** <R>

   prompt: Included angle (+ = ccw, – = cw)<Full circle>

   enter: **360** <R> or <RETURN> for default
4. A blue revolved surface is drawn about the green line.
5. Repeat the REVSURF command three times picking:
   (a) line 2 as path curve then green line as axis of revolution
   (b) line 3 as path curve then green line as axis of revolution
   (c) line 4 as path curve then green line as axis of revolution

Example 1   119

6. From the menu bar select **Draw**
                         3D Surfaces
                         Surface of Revolution
   prompt: Select path curve
   respond: **pick the red polyline**
   prompt: Select axis of revolution
   respond: **pick the green line**
   prompt: Start angle<0> and enter <RETURN>
   prompt: Included angle and enter <RETURN> for 360 degrees.
7. The example demonstrates that:
   (a) both single entities and polyshapes can be revolved
   (b) polyline/arc segments can be used to produce complex surface models
   (c) single line entities are not really suited for REVSURF.
8. Freeze layer OUT and AXIS to remove the original revolved 'shape' and the green
   axis of revolution – Fig. 24.1.
9. Try HIDE and SHADE in each viewport and:
   (a) good effect with the polyline revolved model.
   (b) too 'many lines' with the line shape.
10. Save?

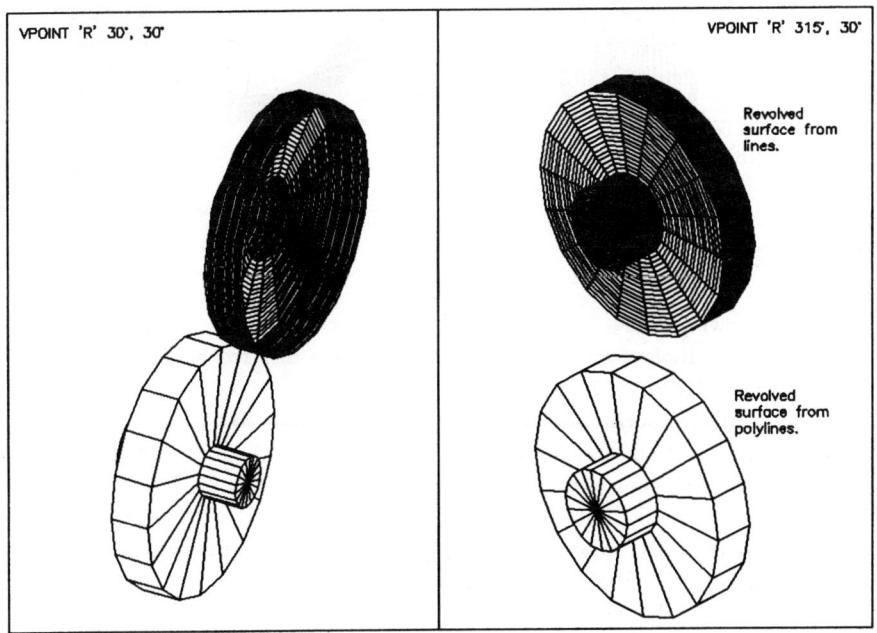

Fig. 24.1   REVSURF example 1 – comparison of LINES and POLYLINES.

EXAMPLE 2

1. Open drawing **EX24_2** from the 3DPACK directory.
2. The screen displays a four viewport configuration with four red polyline shapes and four green lines. Layer SURFACE (magenta) is current.
3. Using the REVSURF command, revolve the red polylines about their respective green axis of revolution for a full circle.
4. HIDE to give Fig. 24.2.
5. Try some other viewpoints then HIDE.

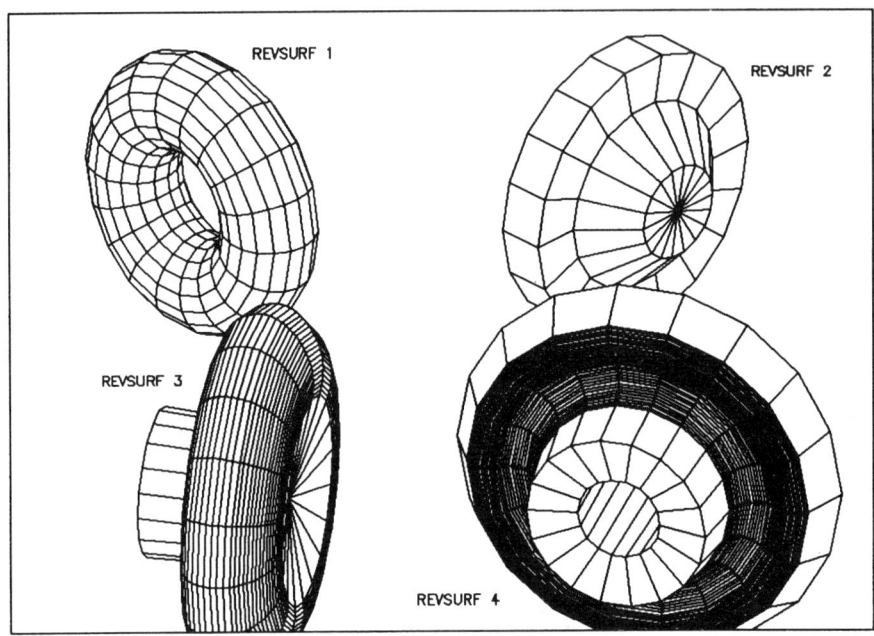

Fig. 24.2   REVSURF example 2 – various polyshapes with HIDEPLOT ON.

EXAMPLE 3

1. Open drawing **EX24_3** from the 3DPACK directory.
2. The screen displays a four viewport configuration of a red polyline and a green axis.Layer SURFACE is current.
3. Revolve the red polyline about the green axis from 0 to 360.
4. HIDE and SHADE in each viewport – interesting result? – there appears to be part of the model missing at the pulley–shaft interface?
5. REGENALL the REDRAWALL.
6. Use the REVSURF command again, pick the polyline as the path curve and the green line as the axis of revolution then:
   (a)   enter 0 as the start angle
   (b)   enter 270 as the included angle.

Example 3   121

7. HIDE, SHADE in each viewport.
8. REGENALL, REDRAWALL.
9. Undo the revsurf effect, or re-open EX24_3.
10. Thaw layer SHAFT and freeze layer SURFACE.
11. With the TOP viewport active, revsurf the blue polyline about the green axis from 0 to 270°.
12. Thaw layer BUSH and freeze layer SHAFT.
13. Revsurf the magenta polyline about the green axis from 0 to 270°.
14. Thaw layers SHAFT and SURFACE.
15. Freeze layers OUT and AXIS.
16. HIDE each viewport – Fig. 24.3.
17. SHADE each viewport.
18. Regenall, etc.

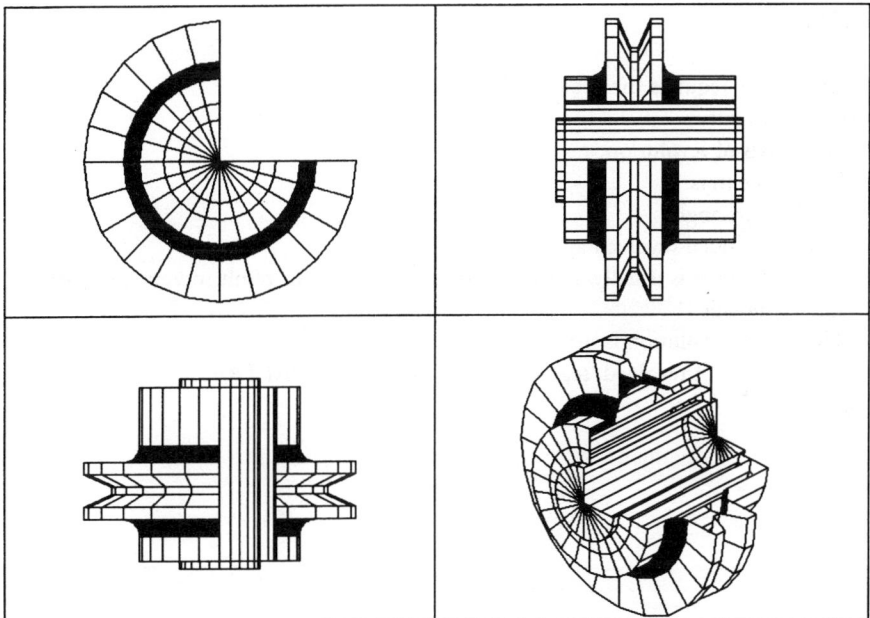

Fig. 24.3   REVSURF example 3 – multiple viewports of PULLEY arrangment.

TASK

1. Erase the revolved surfaces.
2. Repeat the revsurface of the pulley shape, but this time try and pick the green axis at the opposite end of the line from before. Enter angles of 0 and 270° as before. Is there any difference in the ruled surface?
3. Erase the revolved surface, then repeat the command using the pulley polyline with the following angles:
   (a) start 0, included −270
   (b) start 15, included 180
   (c) start 120, included 120
   (d) start −30, included −270

SUMMARY

1. REVSURF requires the user to select:
   (a) a path curve
   (b) an axis of revolution.
2. User also enters:
   (a) the start angle
   (b) the included revolved angle.
3. The path curve is usually a polyline/arc shape. This can result in very complex 3D surface models.
4. The axis of revolution is usually a line.
5. The number of revolved 'segments' is controlled by SURFTAB1.

ACTIVITY

This activity requires that you draw the shape to be revolved, so:

1. Open drawing **ACT_11** from the 3DPACK directory.
2. The screen displays a four viewport configuration with a green axis of revolution. Also displayed (in Paper space) is the outline of a glass with the coordinates of various 'vertices'.
3. With layer OUT current use the PLINE command to draw the glass outline using the coordinates given. The shape consists entirely of line and arc segments. *Note*: you could always design your own glass outline.
4. With layer GLASS current, revolve the outline.
5. HIDE to give the views as Fig. 24.4.

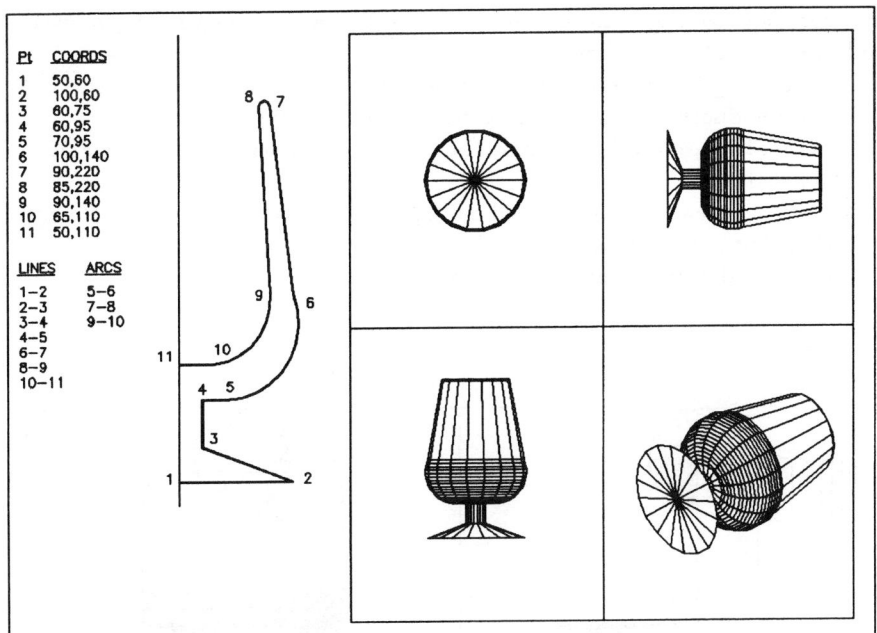

Fig. 24.4   Activity 11. PLINE coordinates for glass REVSURF model.

25

User exercise 3

We have now covered the four surface commands: TAB, EDGE, RULE and REV. To test your understanding of the command, this exercise will require you to add one of each surface to a 3D wire-frame model.

1. Open drawing **USEX3** from the 3DPACK directory.
2. The screen displays a four viewport configuration of the model.
3. Check the layer control dialogue box. There are four new layers, namely TAB, EDGE, RULE and REV.
4. Making each new layer current, add the appropriate surface to the model. No help is given, but it is very simple.
5. Figure 25.1 displays the final expected result.

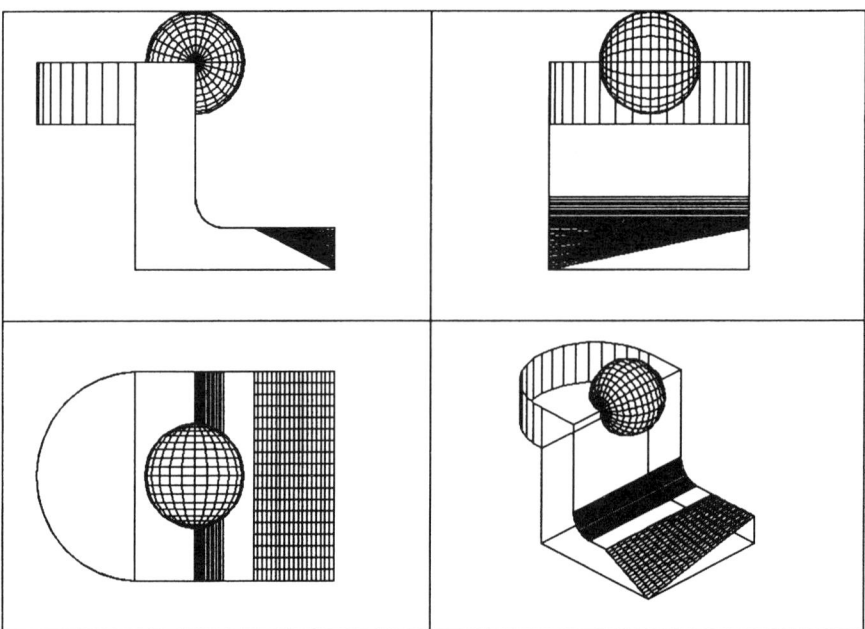

Fig. 25.1 Completed USEX3 with four surfaces added.

26

Three-dimensional objects

AutoCAD has nine predefined 3D objects which can be used in drawings to great effect. The objects are selected and 'inserted' into drawings by the user, who enters various objects sizes and start points.

EXAMPLE 1 – THE NINE OBJECTS

1. Open drawing **EX26_1** from the 3DPACK directory.
2. The screen will display the nine 3D objects available – ten are drawn, as the CONE object is also used for cylinders.
3. Enter HIDE to 'see' the objects in 3D – Fig. 26.1.
4. Try SHADE.
5. The 3D objects are SURFACE models.

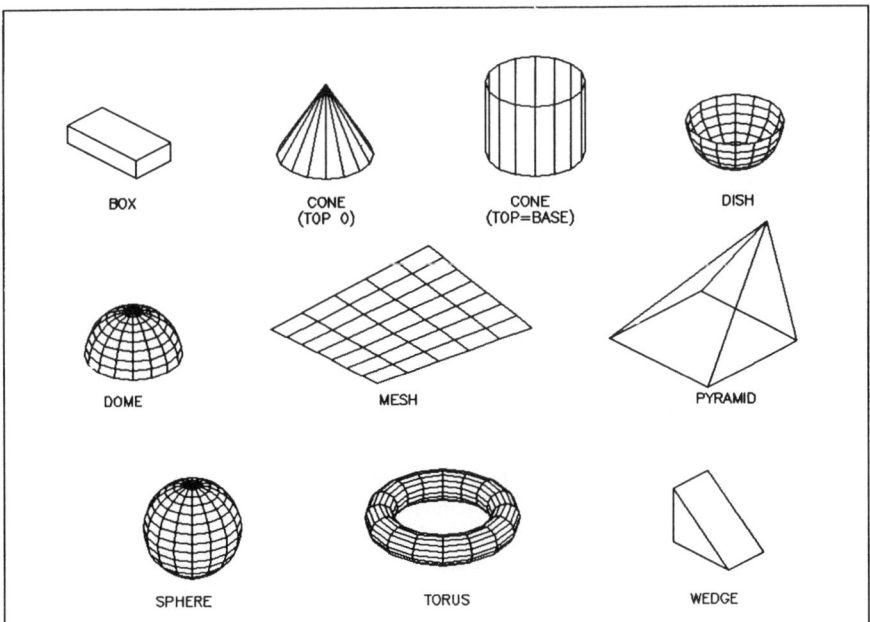

Fig. 26.1 Three-dimensional objects with HIDEPLOT ON.

EXAMPLE 2 – INSERTING 3D OBJECTS INTO A DRAWING

In this exercise, you will do the work, inserting various 3D objects into a drawing. All the relevant prompts and entries will be given. As the exercise progresses, reason out the coordinate input in relation to the object being inserted and the position of the UCS.

1. Open drawing **EX26_2** from the 3DPACK directory.
2. The screen displays a Paper space four viewport configuration with no model. Make layer OBJECTS current and change to Model space. The icon will be in an unusual alignment.
3. From the screen menu select **SURFACES**

   > **3d** or **objects**
   >
   > **Box**

   prompt: Initializing... 3D Objects loaded

   then: Corner of box

   enter: **0,0,0 <R>**

   prompt: Length and enter **100 <R>** – note yellow line

   prompt: Cube/<Width> and enter **80 <R>** – note yellow rectangle

   prompt: Height and enter **60 <R>**

   prompt: Rotation about Z axis and enter **0 <R>**
4. Select from the screen menu **Cone** (we will draw a cylinder) and:

   prompt: Base center point and enter **50,40,60 <R>**

   prompt: Diameter/<radius> of base and enter **40 <R>**

   prompt: Diameter/<radius> of top<0> and enter **40 <R>**

   prompt: Height and enter **30 <R>**

   prompt: Number of segments<16> and **<RETURN>**
5. Repeat the Cone selection and enter:

   Base center point: 50,40,90

   Radius base: 40

   Radius top: 0

   Height: 20

   Segments: 16
6. Using **EDIT–CHPROP**, change the colour of the cylinder and cone to yellow.
7. From the menu bar select **Draw**

   > **3D Surfaces**
   >
   > **3D Objects...**

   prompt: 3D Objects dialogue box

   respond: **pick Dish** from the list and:

   (a) Dish line turns blue

   (b) Dish icon is highlighted

   then: **pick OK**

   prompt: Center of dish and enter **50,40,0 <R>**

   prompt: Diameter/<radius> and enter **40 <R>**

   prompt: Number of longitudinal segments<16> and **<RETURN>**

   prompt: Number of latitudinal segments<16> and **<RETURN>**
8. Change the colour of this dish to green.

Example 2   127

9. Select **SURFACES–3d objects–Wedge** and:
   prompt: Corner of wedge and enter **100,0,0** <R>
   prompt: Length and enter **50** <R>
   prompt: Width and enter **80** <R>
   prompt: Height and enter **60** <R>
   prompt: Rotation about Z axis and enter **0** <R>

10. Repeat the Wedge selection and enter:

Corner:	0,80,0
Length:	20
Width:	80
Height:	60
Rotation:	180

11. Change the colour of both wedges to cyan.

12. Select the Sphere 3d object and enter:

Center:	50,40,130
Radius:	20
Segments:	both 16
Colour:	magenta

13. Finally select the Torus 3d object and enter:

Center:	50,40,–40
Radius torus:	100
Radius tube:	20
Segments:	both 16
Colour:	number 134 (or something else?)

14. Your 3D object drawing is now complete.

15. HIDE in each viewport which should be the same as Fig. 26.2, and you should realize that the 3D objects are all surface models.

16. Try SHADE in each viewport. The effect is quite pleasing, but the sphere and torus 'lack definition'. This effect will be eliminated when we investigate the RENDER command in a later chapter.

17. REGENALL and save your drawing.

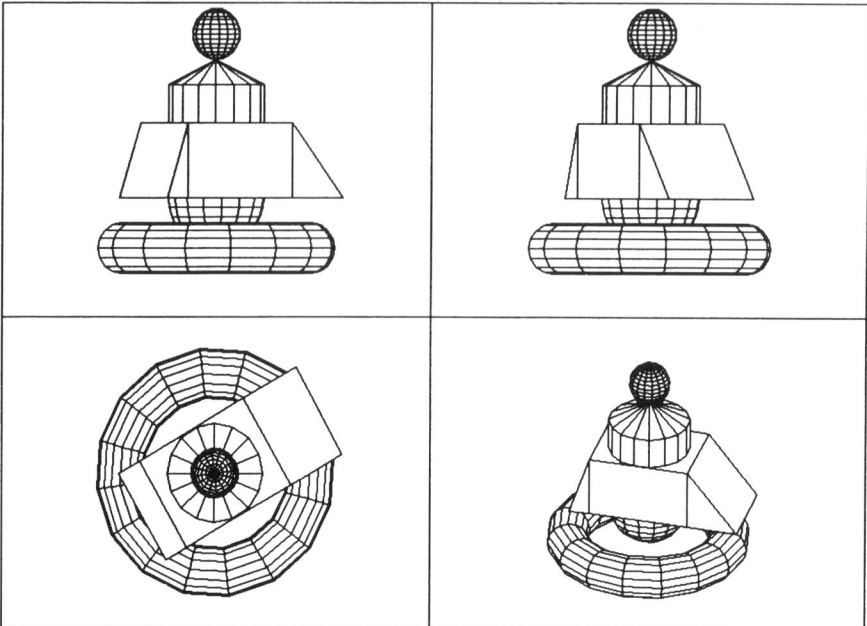

Fig. 26.2   Three-dimensional objects exercise pltted with HIDEPLOT ON.

TASK

Before leaving this section, I would recommend that you investigate the 3D objects command in more detail. Insert objects at various points on the screen, and investigate the rotation about the Z-axis for those objects which have this prompt. A good working knowledge of coordinates and UCS positioning is essential for the centre and corner points.

27

Three 3D commands

In an earlier chapter we investigated how editing in 3D is very dependent on the UCS position and orientation. The commands used in that section (COPY, ARRAY, ROTATE and MIRROR) were 2D commands.

AutoCAD has three commands which are specific to 3D, these being: **Array 3D**, **Mirror 3D** and **Rotate 3D**. We will investigate these commands with worked examples.

EXERCISE 1 – ARRAY 3D

1. Open drawing **EX27_1** from the 3DPACK directory.
2. The screen displays a four-viewport configuration with a red and blue 'box' – both 3D objects. Layer OUT is current and the lower right viewport is active. The UCS is at a base corner of the red box.
3. From the menu bar select **Construct–Array 3D**
   prompt: Select objects
   respond: **window the two objects then <R>**
   prompt: Rectangular or Polar(R/P) and enter **R** <R>
   prompt: Number of rows and enter **5** <R>
   prompt: Number of columns and enter **4** <R>
   prompt: Number of levels and enter **3** <R>
   prompt: Distance between rows and enter **60** <R>
   prompt: Distance between columns and enter **60** <R>
   prompt: Distance between levels and enter **80** <R>
4. The screen will display the rectangular array as Fig. 27.1.
5. In the active viewport, HIDE and SHADE to 'see' the effect.
6. Now either (a) erase all the arrayed objects or (b) re-open EX27_1 again.
7. Activate the Array 3D command again and:
   (a) select the two objects then <R>
   (b) select the Polar (P) option
   prompt: Number of items and enter **7** <R>
   prompt: Angle to fill and enter **360** <R>
   prompt: Rotate objects… and enter **Y** <R>
   prompt: Centre point of array and enter **30,120,0** <R>
   prompt: Second point on axis of rotation and enter **100,100,100** <R>
8. The two objects will be polar rotated about the two entered points.
9. Use HIDE to 'see' the effect – Fig. 27.2.

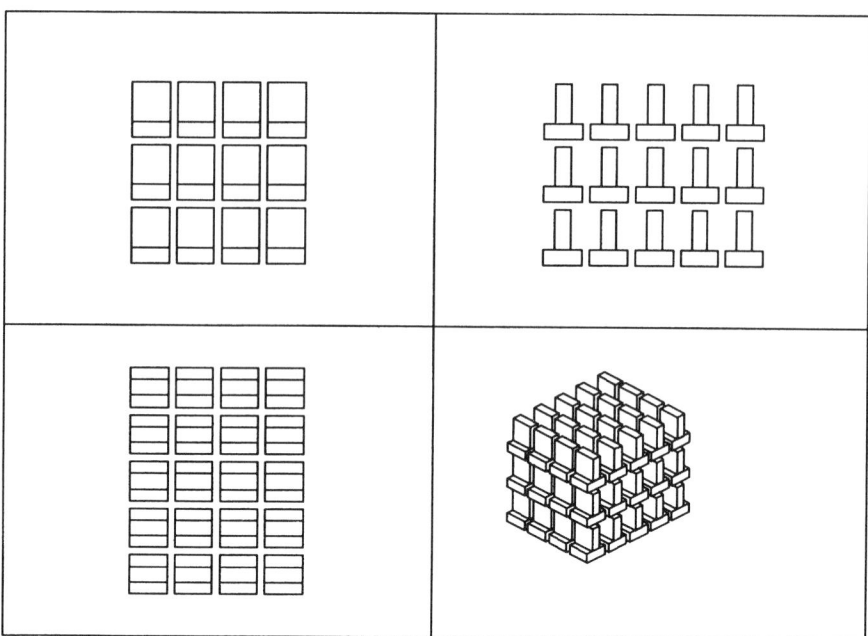

Fig. 27.1 Rectangular 3D array with UCS BASE.

Fig. 27.2 Polar 3D array with UCS base.

TASK

The original EX27_1 has three saved UCS positions: BASE, POSA and POSB. Investigate both the rectangular and polar 3D array options using the POSA and POSB UCS positions. Use the information given in the worked example and compare the results. Try and reason out the final result.

EXERCISE 2 – MIRROR 3D

1. Open drawing **EX27_2** from the 3DPACK directory.
2. The screen displays a four-viewport configuration with:
   (a) layer OUT current and the lower right viewport active
   (b) a red box and a green wedge on top of the box
   (c) black numbers for reference.
3. From the menu bar select **Construct–Mirror 3D**
   prompt: Select objects and pick the wedge then <R>
   prompt: Plane by Entity...<3 points>
   respond: MIDpoint and pick line 1
   prompt: 2nd point on plane – MIDpoint and pick line 2
   prompt: 3rd point on plane – MIDpoint and pick line 3
   prompt: Delete old objects and enter N <R>
4. The green wedge will be mirrored about the top 'left side' of the box.
5. Repeat the Mirror 3D command and:
   prompt: Select objects and pick the two green wedges then <R>
   prompt: Plane by Entity...
   enter: **YZ** <R>
   prompt: Point on *YZ* plane (0,0,0)
   respond: MIDpoint and pick line 4
   prompt: Delete old objects and enter N <R>
6. The two green wedges will be mirrored about the top surface of the box – Fig. 27.3.
7. At this stage SAVE the drawing, as it will be used in the next exercise on Rotate 3D
   – you should be able to manage this.
8. HIDE and SHADE.

TASK

Erase the mirrored wedges. Can you use the Array 3D command to obtain the same effect as the two mirrored operations. The red box has dimensions of 100 × 100 × 30.

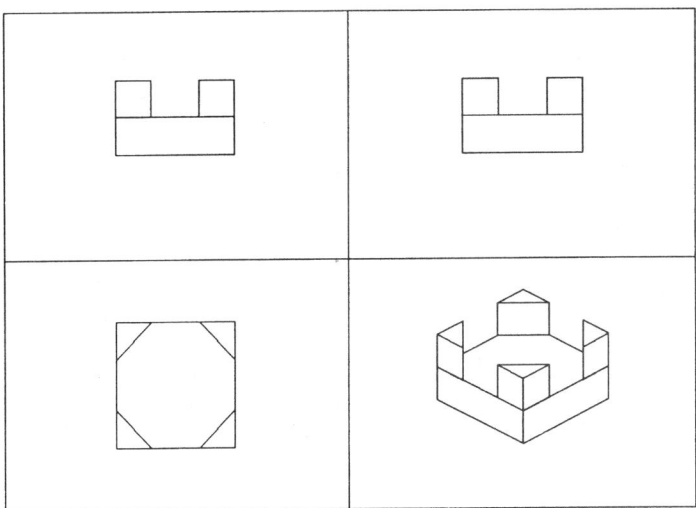

Fig. 27.3   Mirror 3D command exercise.

EXERCISE 3 – ROTATE 3D

1. Open the model saved from the Mirrored exercise.
2. From the menu bar select **Modify**

    Rotate 3D

    prompt: Select objects
    respond: **pick the ORIGINAL green wedge then <R>**
    prompt: Axis by Entity…
    enter: **X <R>**
    prompt: Point on X axis(0,0,0)
    enter: **0,0,30 <R>**
    prompt: <Rotation angle>/Reference
    enter: **90 <R>**
3. Repeat the 3D Rotate command, select the same wedge then:
    prompt: Axis by Entity…
    enter: **Z**
    prompt: Point on Z axis and enter **0,0,30<R>**
    prompt: Rotation angle and enter **−90<R>**
4. Now MOVE the rotated wedge from the point **0,0,30** by **@30,0,−30**

TASK

Can you 3D Rotate and Move the other three wedges to give Fig. 27.4? Each wedge requires two rotations (?) and the wedge dimensions are 30 × 30 × 30. Use OSNAP END(?) to select reference points for the move command.

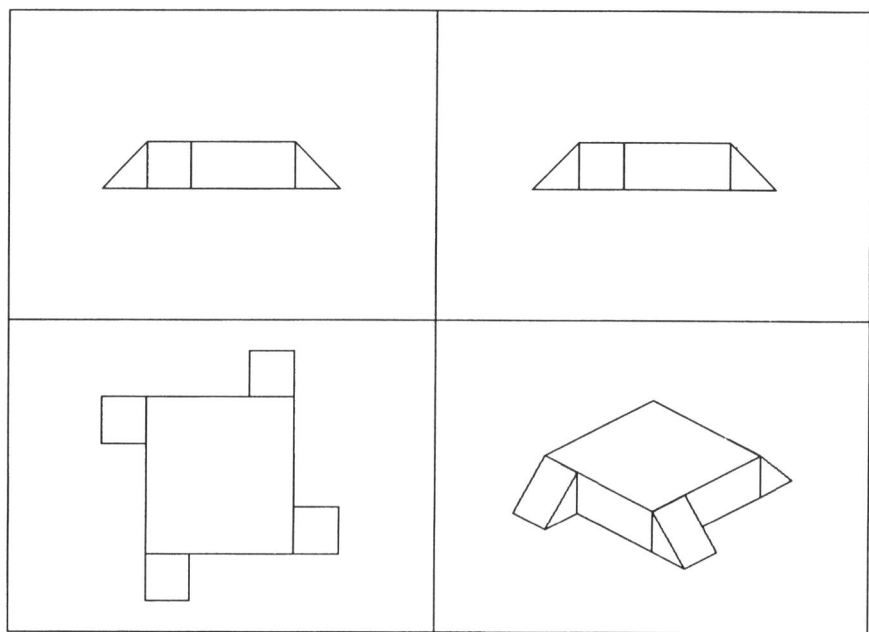

Fig. 27.4   Rotate 3D command exercise.

SUMMARY

- ARRAY 3D – can be rectangular or polar
    - rectangular arrays have (a) rows in the Y-direction
                                 (b) columns in the X-direction
                                 (c) levels in the Z-direction
    - polar arrays are about a centre axis. This axis can be entered as co-ordinates or selected by referencing existing entities.
- MIRROR 3D – can be (a) about three points selected by the user
                         (b) about a plane
                         (c) about the Z-axis
                         (d) about an entity.
- ROTATE 3D – is (a) about an axis
                     (b) about an entity.

28

Blocks in 3D

Blocks are employed when frequently used objects have to be inserted into the current drawing. Blocks of 3D models can be inserted in a similar manner to 2D blocks. Our example will use three blocks created from 3D objects.

1. Open drawing **EX28_1** from the 3DPACK directory and:
    (a) screen displays a three-viewport configuration
    (b) layer OBJECTS is current
    (c) no model is displayed.
2. From the screen menu select **BLOCKS**

    BLOCK

    ?

    prompt: Blocks to list<*> and enter <R>
    prompt: Text screen with:
    Defined blocks
    CYL
    RECT
    WED
    User
    Blocks
    3
3. Flip back to the drawing screen with F1.
4. From the menu bar select **Draw**

    Insert...

    prompt: (a) cancel the X at the Specify Parameters on Screen by picking the box to remove the X
    (b) pick **Block...** to display the Blocks Defined in this Drawing dialogue box.
    (c) pick **RECT** (turns blue)
    (d) pick OK to return to the Insert dialogue box
    (e) enter the following values:

Insertion pt	*Scale*	*Rotation*
X: 0	X: 1	Angle: 0
Y: 0	Y: 1	
Z: 0	Z: 1	

    (f) pick OK.

5. A blue box shape will be inserted at the point 0,0,0.
6. Repeat the **Draw–Insert...** selection and:
   (a) ensure RECT is the block name – it should be
   (b) ensure Specify parameters box not active, i.e. no X
   (c) enter the following values:

Insertion pt	Scale	Rotation
X: 0	X: 1	Angle: 0
Y: 30	Y: 0.5	
Z: 30	Z: 1	

   (d) pick OK.
7. **Draw–Insert...** again and:
   (a) pick Block... then CYL and OK
   (b) change the following:

Insertion pt	Scale	Rotation
X: 15	X: 1	Angle: 0
Y: 15	Y: 1	
Z: 30	Z: 1	

8. Repeat the CYL insertion another five times, changing the insertion points as follows:
   (a) 75,15,30   (b)   135,15,30
   (c) 15,105,30   (d)   75,105,30 (e)   135,105,30
   The scale factor is 1 and the rotation angle is 0.
9. Insert the block RECT with the following values:
   (a)

Insertion pt	Scale	Rotation
X: 0	X: 1	Angle: 0
Y: 0	Y: 1	
Z: 130	Z: 1	

   (b)

Insertion pt	Scale	Rotation
X: 50	X: 0.3333333	Angle: 0
Y: 40	Y: 0.3333333	
Z: 160	Z: 3	

10. Use the **Draw–Insert...** sequence, select the **WED** block and insert it four times using the following:
    (a)

Insertion pt	Scale	Rotation
X: 100	X: 1	Angle: 0
Y: 40	Y: 1	
Z: 160	Z: 1	

    (b)

Insertion pt	Scale	Rotation
X: 50	X: 0.8	Angle: –90
Y: 40	Y: 1.125	
Z: 160	Z: 1	

    (c)

Insertion pt	Scale	Rotation
X: 100	X: 0.8	Angle: 90
Y: 80	Y: 1.125	
Z: 160	Z: 1	

    (d)

Insertion pt	Scale	Rotation
X: 50	X: 1	Angle: 180
Y: 80	Y: 1	
Z: 160	Z: 1	

11. If you have managed all the insertions, your screen now displays a temple (of sorts). This is a 3D model made from 3D blocks.
12. Hide to give Fig. 28.1.

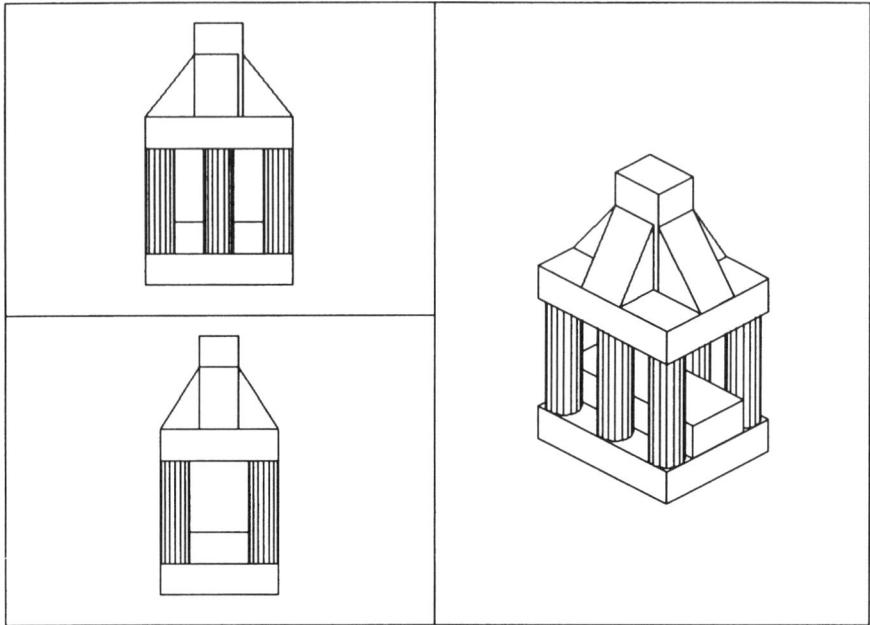

Fig. 28.1   Completed 3D BLOCK exercise.

TASK

Can you change the colour of the blue inserted blocks?

SUMMARY

1. Three-dimensional blocks are created and inserted into a drawing in a manner similar to 2D blocks.
2. The Insert dialogue box is recommended, as it allows all parameters to be entered at the one time. The Specify Parameters on Screen option should be cancelled.
3. When a 3D block is created and the insertion point is not on the *XY*-plane, a warning is displayed for the user.

ACTIVITY

I have not added any activity for this chapter, but the user should try and create some 3D blocks of their own, and insert them into a drawing.

29

Three-dimensional Wblocks

All drawings are Wblocks and all Wblocks are drawings.

I'll leave the user to ponder the statement above and decide if it is true or not. Three-dimensional Wblocks can be used in a similar manner to 2D Wblocks, the only difficulty being to decide on the insertion point relative to the UCS position and orientation.

THREE-DIMENSIONAL WBLOCK EXERCISE

This exercise should be quite interesting to the user, as it involves a 3D chessboard created from two coloured 'squares'. I have also included two of the chess pieces (pawn and rook) to be positioned on the board.

The board

1. Open drawing **EX29_1** from the 3DPACK directory.
2. The screen displays a single viewport in 3D configuration. The UCS icon is at the centre-left of the screen.
3. From the screen menu select **BLOCKS**
                              INSERT
                              File...
   prompt: Select Drawing File dialogue box
   respond: (a) double left click on \
            (b) double left click on 3DPACK
            (c) scroll down and pick SQUARE2 – turns blue
            (d) pick OK – note 'ghost' image
   prompt: Insertion point and enter **0,0,0** <R>
   prompt: *X* scale and <R> for full size
   prompt: *Y* scale and <R> for full size
   prompt: Rotation angle and enter **−90** <R>
4. A yellow 'square' is displayed at the insertion point. *Note*: the −90 rotation angle is necessary due to the way in which I created the WBLOCK.

5. From the menu bar select **Draw–Insert...**
   prompt: Insert dialogue box
   respond: (a) cancel the X at Specify Parameters on Screen
             (b) pick File...
             (c) ensure 3DPACK is current directory – it should be
             (d) scroll down and pick SQUARE1
             (e) pick OK – returns to Insert dialogue box
             (f) enter the following:

Insertion pt	Scale	Rotation
X: 60	X: 1	Angle: −90
Y: 0	Y: 1	
Z: 0	Z: 1	

6. A magenta 'square' is displayed alongside the yellow square.
7. Using the screen menu repeat the INSERT command using the following:
   (a) SQUARE1 (magenta) at 0,60,0 full size at −90 rotation
   (b) SQUARE2 (yellow) at 60,60,0 full size at -90 rotation.
   Note: don't panic if there is a redefine message, just choose no.
8. Multiple COPY the four squares from a base point of 0,0,0 by:
   (a) @120,0,0
   (b) @240,0,0
   (c) @360,0,0.
9. Use the ARRAY command (not Array 3D), select all 16 squares and enter the
   following:
   (a) R for rectangular
   (b) 4 as number of rows
   (c) 1 as number of columns
   (d) 120 as the row distance.
10. The 64 square chessboard will be drawn as alternate yellow and magenta squares.
11. HIDE, SHADE then REGEN – nice effect?
Note: the chessboard could have been created solely using the INSERT command. I
thought that using COPY and ARRAY would break the monotony of inserting at various
points.

The chess pieces

1. Using **Insert–File...** pick PAWN from the Select Drawing File list and:
   (a) insertion point: 30,90,10
   (b) *X* and *Y* scale: 1, i.e. full size
   (c) rotation: 0.
2. A red pawn is displayed at the insertion point.
3. Repeat the PAWN insertion with:
   (a) insertion point: 30,390,10
   (b) X and Y scale: 1
   (c) rotation: 0.
4. Use CHPROP to change the colour of the second inserted pawn to blue. You may
   have to explode first?
5. Now insert the piece called ROOK at the point 30,30,10 full size with 0 rotation.

Title block

Before you begin your task, enter Paper space and insert the Wblock TITLE from the 3DPACK directory with:
   (a) insertion point: 250,20
   (b) full size with 0 rotation.

TASK

You now have to:
   (a) display all eight red and blue pawns
   (b) display the other three rooks (one red and two blue).
There are various options to achieve this:
   (a) inserting
   (b) copy
   (c) array.
For your information the SQUARES have dimensions of 60 × 60 × 10.

   When complete, your drawing should resemble Fig. 29.1.

Fig. 29.1 CHESSBOARD created from 3D WBLOCKS.

SUMMARY

Three-dimensional WBLOCKS are inserted into a drawing in a manner similar to 2D WBLOCKS.

Additional activity

On a dark, rainy night when you have nothing to do, try and add the other remaining chess pieces to your board.

As a starting point, I created the PAWN and ROOK using the REVSURF command, creating a 2D polyline and revolving it about the *y*-axis. The final model was then 3D rotated by 90° to 'stand' on the *XY*-plane, i.e. vertically upwards in the *Z*-direction. The WBLOCK was created selecting the CENTRE of the base circle as the insertion point. All my chess pieces were created with a 20-radius base. The height of the piece is arbitrary, but obviously depends on the actual piece being made.

I hope this helps if you decide to proceed with this drawing.

30

Attributes in 3D

Attributes allow information (in the form of text) to be added to blocks. While basically a 2D facility, it is possible to add attributes to 3D models. As the previous two sections have been about blocks, this seems a reasonable time to investigate attributes.

ATTRIBUTE EXERCISE

The exercise to be used to demonstrate attributes with 3D models will be a machine shop, the machines being represented by 3D blocks. Each block will contain certain information about the machine, e.g. the type of machine, its manufacturer, the cost and the overall dimensions. This information will be 'attached' to the block as attributes.

1. Open drawing **EX30_1** from the 3DPACK directory.
2. The screen displays:
   (a) a three-viewport configuration – 3D view, top view and front view of the machine shop
   (b) red lines indicated the machine shop size and 'walkways'
   (c) a blue machine in position
   (d) no attribute information?
3. From the screen menu select **DISPLAY**

                    ATTDISP – attribute display

   prompt: Normal/ON/OFF<Off>

   enter: **ON** <R>
4. Four blue attributes are displayed on the top of the blue machine in each viewport. These attributes give:
   (a) the machine type: TURNING CENTRE
   (b) the manufacturer: BOBMACH
   (c) the cost: 250000
   (d) the overall machine dimensions: $60 \times 100 \times 50$

INSERTING BLOCKS WITH ATTRIBUTES

1. Select from the screen menu **BLOCKS–INSERT**
   prompt:　Block name and enter **M2 <R>**
   prompt:　Insertion point and enter **40,200,0 <R>**
   prompt:　*X* and *Y* scale and **<RETURN>** for full size
   prompt:　Rotation angle and enter **0 <R>**
   prompt:　Enter Attributes dialogue box displaying :-
   　　　　(a)　Block name: M2
   　　　　(b)　Attribute prompts, e.g. type of machine, etc.
   　　　　(c)　Default values, e.g. aaaa
   respond:　alter the default values to:
   　　　　1. Type of machine: MACHINING CENTRE
   　　　　2. Machine manufacturer: MACHCO
   　　　　3. Price of machine: 570000
   　　　　4. Machine dimensions: 40 × 80 × 80
   　　　　5. pick OK.
2. A yellow block with yellow attribute text is displayed at the insertion point. *Note*: the yellow text may be difficult to 'see' on a white background. Change colour?
3. Repeat the INSERT command with the following information:

Block name:	M3	M4
Insertion point:	160,15,0	160,145,0
Type of machine:	AUTOMAN	HORIZ CENTRE
Machine manufacturer:	MANTAG	BOBMACH
Price of machine:	678000	1250000
Machine dimensions:	80 × 100 × 70	60 × 120 × 50

4. Figure 30.1 displays the result of the insertions.

Notes

1. The blocks were created from 3D objects – box.
2. The attributes are positioned 2mm above the top of each box.

USING ATTRIBUTES

Attribute information can be *extracted* from a drawing, and used as input to other computer packages, e.g. databases, spreadsheets, etc.

The following sequence will extract the attribute information from the four blocks in the machine shop layout drawing.

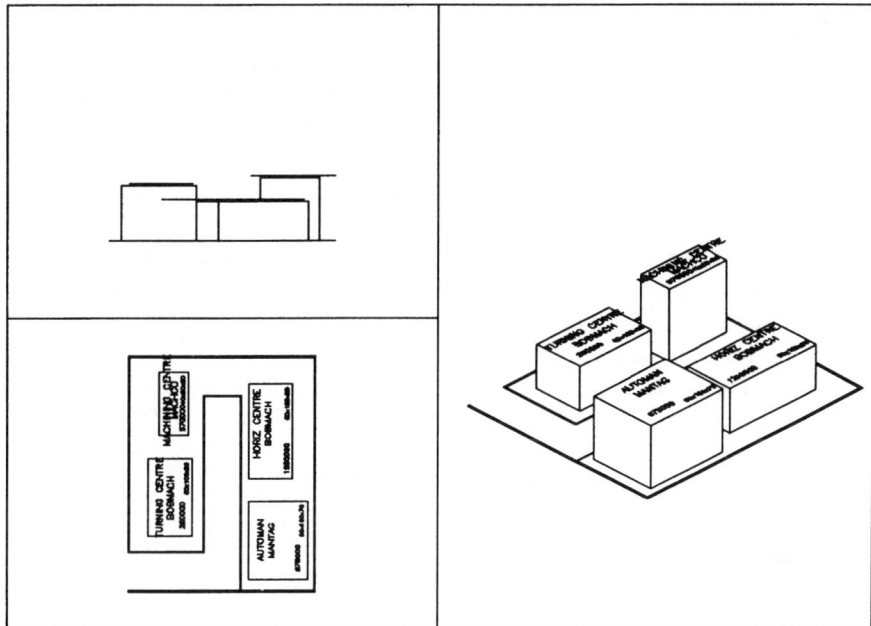

Fig. 30.1 Machine shop layout with blocks and attributes added.

1. From the screen menu select **UTILITY**

                        ATTEXT – attribute extraction

    prompt: CDF, SDF or DXF Attribute extract

    enter: **S** <R> – for SDF format

    prompt: Select Template File dialogue box

    respond: (a)   ensure 3DPACK is directory name

                (b)   pick MACHINE from file list – turns blue

                (c)   pick OK

    prompt: Create extract file dialogue box

    respond: **pick type it**

    prompt: Command line

    and: Extract file name<?>

    enter: **C:\3DPACK\SHOP** <R>

    prompt: 4 records in extract file

    and: command line returned.

2. At the command line enter **TYPE** <R>

    prompt: Text screen

    and: File to list

    enter: **C:\3DPACK\SHOP.TXT** <R>

3. The screen will display attribute information for the four machine blocks in the drawing. You may recognize the information which is displayed, e.g. the block name, the X- and Y-insertion points, the various attribute information which you entered.

Notes

1. It was not my intention to include attributes in this pack. This chapter has only been an introduction to how attributes can be added and extracted from 3D blocks. Attributes can be edited and extracted in different formats, but these topics are beyond the scope of this book.
2. The extraction of attributes requires that two different TEXT type files are used. These are:
   (a) a **TEMPLATE** file which contains the 'program' to extract the attribute information from the drawing. This is usually written by the user and the file is usually in the same directory as the drawing. In our example the template file was MACHINE.TXT, and is listed in Fig. 30.2
   (b) an **EXTRACT** file which contains the actual extracted attribute information. Our extraction file was SHOP.TXT and its contents are displayed in Fig. 30.3.

```
BL: NAME      C003000
BL:X          N006001
BL:Y          N006001
TYPE          C018000
MAKER         C008000
COST          N008000
SIZE          C011000
```

Fig. 30.2 Template text file: MACHINE.TXT.

```
M1  25.0    70.0TURNING CENTRE     BOBMACH  25000060x100x50
M2  40.0   200.0MACHINING CENTRE MACHCO  57000040x80x80
M3  160.0   15.0AUTOMAN            MANTAG  67800080x100x70
M4  160.0  145.0HORIZ CENTRE       BOBMACH  125000060x120x50
```

Fig. 30.3 Extract text file: SHOP.TXT.

31

Viewport specific layers

AutoCAD has the facility for layers which are specific to individual viewports. The concept will be illustrated by adding dimensions to a multi-view drawing of a 3D model.

EXERCISE 1 – USING VIEWPORT SPECIFIC LAYERS

1. Open drawing **EX31_1** from the 3DPACK directory.
2. The screen displays a four viewport configuration with:
   (a) two red 3D objects – a box and a wedge
   (b) lower right viewport active and layer DIMENS (magenta) current
   (c) two dimensions (150 and 100) displayed in all viewports
   (d) black donuts and numbers for reference.
3. Restore the UCS FRONT.
4. With the lower right viewport active, vertical dimension lines 1–2 and 2–3 using the continuous option.
5. The dimensions are added to the lower right viewport *and also to the other three viewports*.
6. Figure 31.1 displays the added dimensions.
7. Activate the Layer Control dialogue box and note the four layers DIMBL, DIMBR, DIMTL, DIMTR all have On . . C N as the State. These are viewport specific layers.
8. Cancel the dialogue box, and make the lower left viewport active.
9. With the Layer Control dialogue box:
   (a) pick **DIMBL** – turns blue and other options become available
   (b) pick **Cur VP: Thw** – the C disappears at the layer line
   (c) pick **Current**
   (d) Freeze layer DIMENS
   (e) pick OK.
10. The lower left viewport will display two green dimensions and:
    (a) no magenta dimensions are displayed – DIMENS frozen
    (b) DIMBL is the layer name in the Status line.
11. Make the upper left viewport active and with the Layer Control dialogue box:
    (a) pick DIMTL
    (b) pick Cur VP: Thw
    (c) pick Current then OK.
12. Three green dimensions will be displayed in the active viewport.

13. With the top right viewport active, use the Layer Control dialogue box to make DIMTR the current layer, remembering to **CurVP: Thw** it first.
14. The screen display now has individual dimensions in three of the viewports – Fig. 31.2.
15. Finally make the lower-right viewport active and with the layer control dialogue box, CurVP: Thw layers DIMBL, DIMTR and DIMTL. The lower-right viewport should display the seven green dimensions and is a bit of a mess!

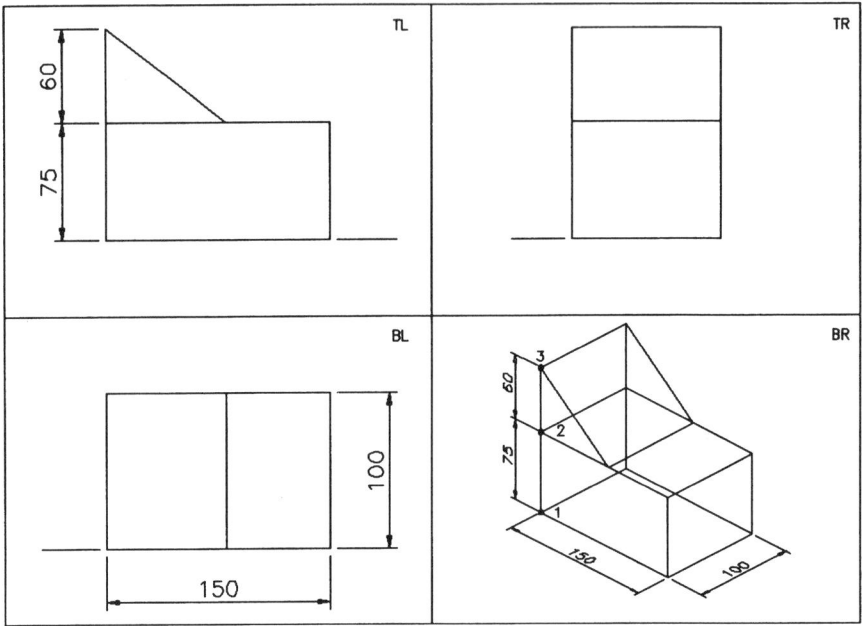

Fig. 31.1 Dimensions added with layer DIMEWSN current.

VIEWPORT SPECIFIC LAYERS

When layers are made they are **GLOBAL**. This means that if a multi-view model is created, what is drawn in one viewport will be displayed in all the others – as happened with the magenta dimensions on layer DIMENS. Viewport specific layers allow entities to be created on layers which are specific to an individual viewport. This allows dimensions (and other entities) to be displayed in one viewport only. The entities are also created in the other viewports, but the specific layers are frozen in these viewports.

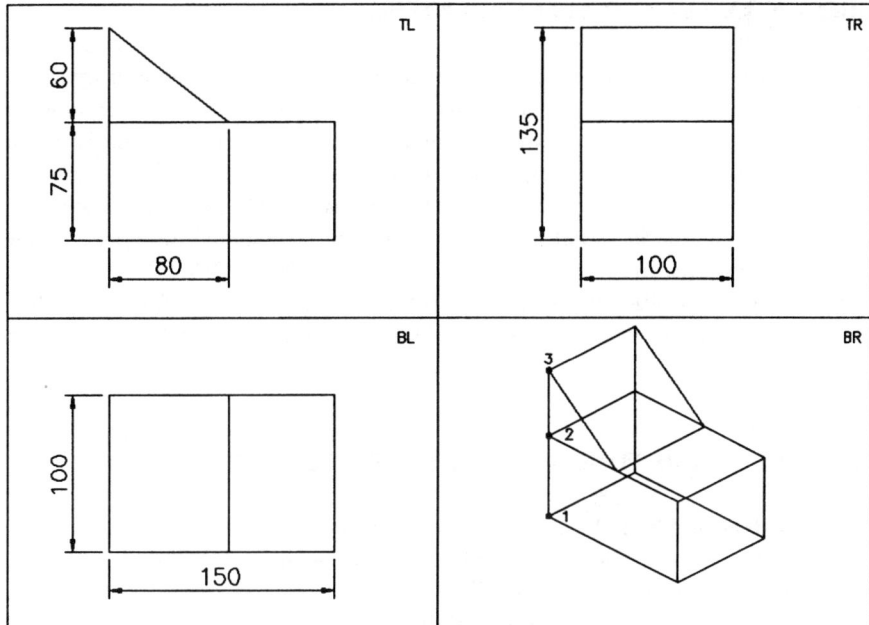

Fig. 31.2 Model with dimensions added using viewport specific layers.

LAYER STATE

Most users should be familiar with the Layer Control dialogue box which gives information about the state of layers, i.e. whether they are on, frozen, locked, etc. The state of specific viewport layers is also displayed in this dialogue box, and the following details the letters which are used:

On . . . .   layer ON
On F . . .   layer Frozen
On . L . .   layer Locked
On . . C N   New viewport specific layer frozen

'Normal' layers can have several states:

(a) ON or OFF
(b) THAWED or FROZEN (F)
(c) UNLOCKED or LOCKED (L).

Viewport specific layers can be:

(a) THAWED or FROZEN and C signifies the layer is frozen
(b) NEW THAWED or FROZEN and N signifies newfreeze.

CREATING VIEWPORT SPECIFIC LAYERS

Viewport specific layers are created by the user in a manner similar to ordinary layers, but a different command is used. It is usual to create a viewport specific layer for every viewport, i.e. if there are four viewports there should be four specific viewport layers for dimensions and other entities – especially true when solid modelling. For our purpose we will only consider viewport specific layers for dimensions. The names which are given to these viewport specific layers should be meaningful and should easily identify the individual viewports. In our example the names were DIMBL, DIMTR, etc., and I'm sure that you have realized that these mean bottom left and top right. Another method of identifying viewport layers is to use HANDLES which will be considered in our next exercise.

EXERCISE 2 – CREATING VIEWPORT SPECIFIC LAYERS

1. Open drawing **EX31_2** from the 3DPACK directory.
2. The screen displays a four viewport configuration with four 3D objects.
3. From the screen menu select **SETTINGS**
   HANDLES

   prompt: Handles are disabled
   then: ON/DESTROY
   enter: **ON** <R> or pick ON from the screen menu.
   Note: your prompt may display 'Handles are enabled. Next handle.' If it does, don't worry. Enter ON <R> and the prompt will then display 'Handles are already on'.
4. Enter Paper space and select from the screen menu **AutoCAD**
   INQUIRY
LIST

   prompt: Select objects
   respond: pick the yellow border of top right viewport then <R>
   prompt: Text screen
   with: VIEWPORT Layer: VP
   Space: Paper space

   Handle = 6
   Status: On and Active
Scale relative to Paper space: 0.5000xp

   Center point, X = 280.00, Y = 197.50, Z = 0.00
   Width 180.00
   Height 125.00
5. Repeat the LIST command and select the other yellow viewport border, noting the handle number. Hopefully you will have:
   Top right: 6
   Top left: 7
   Bottom left: 8
   Bottom right: 9

6. Now select from the menu bar **View**

           Mview

           Vplayer

prompt: ?/Freeze/Thaw/Reset/Newfrz/Vpvisdflt
enter: N <R> – for New freeze
prompt: New Viewport frozen layer name(s)
enter: **DIM_6,DIM_7,DIM_8,DIM_9 <R>**
prompt: ?/Freeze/Thaw...
enter: F <R> – for Freeze
prompt: Layer(s) to Freeze
enter: **DIM_6,DIM_7,DIM_8,DIM_9 <R>**
prompt: All/Select/<Current>
enter: A <R> – for All
prompt: ?/Freeze/Thaw...
enter: <RETURN> to end sequence.

Note: the viewport specific layers which we have made have used the handles to identify them. We could have used another set of names similar to the first exercise, i.e. DIMBL, DIMTR, DIMBR and DIMTL.

7. Activate the Layer Control dialogue box and note the four new layer names which have been added. They are displayed with a C and an N at each line and:

   C: viewport specific layer frozen

   N: new viewport specific layer.

8. Select the four new layers and change their colour to magenta then pick OK.

9. Enter Model Space and make the lower left viewport active. This viewport has handle 8 and we will use layer DIM_8 as its viewport specific layer.

10. Activate the layer control dialogue box and:

    (a) pick **DIM_8**

    (b) pick **Cur VP: Thw**

    (c) pick **Current**

    (d) pick **OK.**

11. Now horizontal and vertical dimension as shown in Fig. 31.3 using the continuous option.

12. Your added dimensions are probably quite small, so select from the screen menu

    DIM

    Dim Vars

    next

    dimscale

    prompt: Current Value<1.00> New value
    enter: 1.5 <R>
    prompt: Dim
    respond: pick (a) DimMenu
                    (b) Edit
                    (c) Update
    prompt: Select objects
    respond: **pick the 5 dimensions then <R>**

13. The added dimensions should be 'clearer'.
14. Make the upper left viewport active, then using the Layer Control dialogue box:
    (a) pick DIM_7
    (b) pick Cur VP: Thw
    (c) pick Current then OK.
15. Change the UCS setting with the sequence UCS <R>
    X <R>
    90 <R>
    UCS <R>
    S <R>
    FRONT <R>
16. Now add the dimensions as shown in Fig. 31.3.

TASK

Add dimensions to the top right viewport (handle 6). You will need to thaw the viewport specific layer and make it current. You will also have to 'reset' the UCS with a rotation about the *Y*-axis. Your final drawing should be similar to Fig. 31.3.

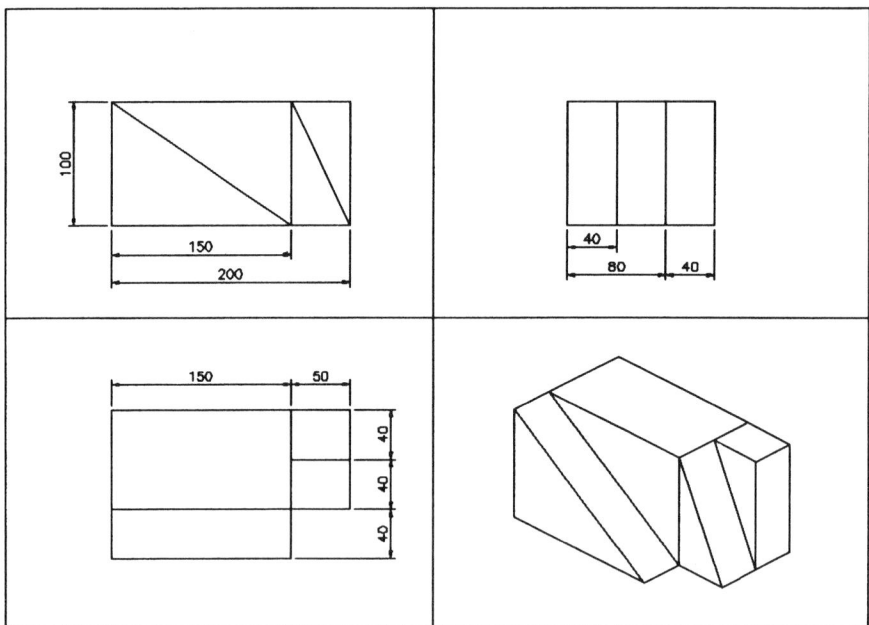

Fig. 31.3   Viewport specific layer exercise with all dimensions added.

SUMMARY

1. Layers can be made which are specific to individual viewports.
2. The command is VPLAYER.
3. Normal layers are GLOBAL.
4. Viewport specific layers can be thawed (Thw) or frozen (Frz).
5. The letter C is used to designate a frozen viewport specific layer.

ACTIVITY

It is now some time since you have had an activity to attempt. Open **ACT_12** from the
3DPACK directory. The screen will display a four viewport configuration with a red wire-
frame model. A bit of RULESURF has been added.

You have to make four viewport specific layers and add the dimensions as shown in
Fig. 31.4. Several UCS settings have been made to assist with the task. These will need
to be used correctly.

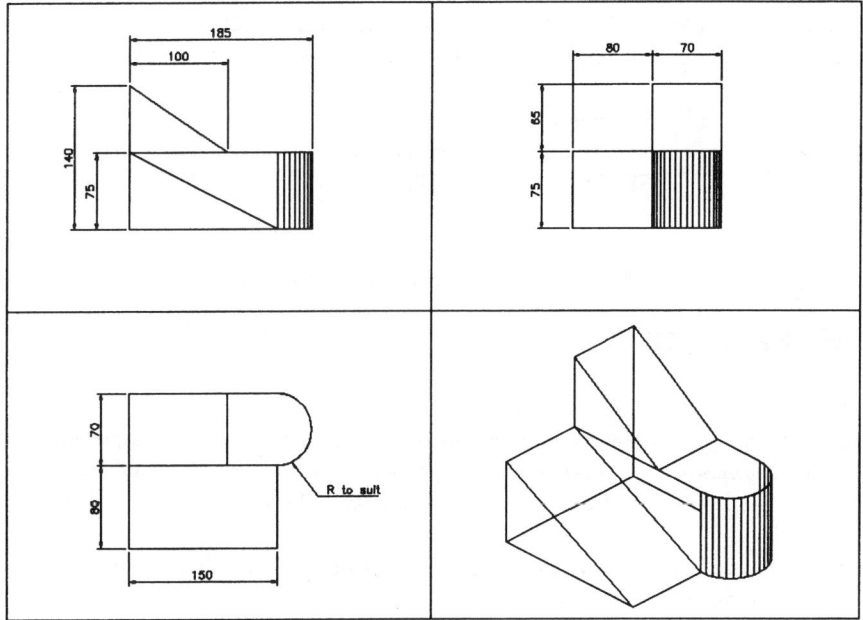

Fig. 31.4 Activity 12 – dimensioning with viewport specific layers.

32

Dynamic viewing

Three-dimensional models can be viewed from different angles using the VPOINT command. A very useful addition to viewing 3D models is the dynamic view (DVIEW) which allows the user a variety of options, and introduces 'perspective' to models. The command is 'interactive' in that several options can be used one after the other and undone as required.

The concept of dynamic viewing is that you (the viewer) have a **CAmera** which is looking at the object (the **TArget**). The target is a certain **Distance** from the camera, and the camera can be **POinted** to a specific point on the target. It is possible to **PAn** and **Zoom** the target, and **CLip** parts of the target away. It is possible to **Hide** the target at any time.

The command can be activated in three ways:

Menu bar	*Screen menu*	*Command line*
View	DISPLAY	enter DVIEW <R>
Set View	DVIEW	
Dview		

We will investigate the command with two worked exercises.

DVIEW EXERCISE 1

1. Open drawing **EX32_1** from the 3DPACK directory.
2. A six viewport configuration of a 'sort of' red wire-frame model is displayed. The top left viewport (a) is active. All viewports are at VPOINT R 315 and 30.
3. Use the layer control dialogue box to thaw layers SURF1, SURF2, SURF3 and SURF4.
4. HIDE and SHADE in viewport (a) to 'see' the model in detail, then REGEN. The shaded model is quite nice?
5. Make viewport (b) active.
6. From the screen menu select **DISPLAY**
   DVIEW

   prompt: Select objects
   respond: **window the complete model then <R>**
   prompt: CAmera/TArget/Distance/Points/PAn/Zoom/TWist/CLip/Hide/
   Off/Undo/<eXit>

   Note: this is the normal DVIEW prompt line. The various options are activate by entering the capital letters, e.g. CA,TW,CL,H, etc.

enter: **CA <R>** – camera option

and: 'ghost' image of model appears and moves as the mouse is moved. The status lines displays an Angle value, which changes as the mouse is moved.

prompt: Toggle angle in/Enter angle from *XY*-plane

enter: **–40 <R>**

prompt: Toggle angle from/Enter angle in *XY*-plane from *X*-axis

enter: **–30 <R>**

prompt: CAmera/TArget...

enter: **H <R>** – hide option to 'see' result of camera entries

prompt: CAmera...

enter: **X <R>** to end command.

7. The model will be displayed with the camera option and the angle values entered.

8. With viewport (c) active, use DVIEW and window the complete model then <R>

prompt: CAmera/TArget...

enter: **TA <R>** – target option

prompt: Toggle angle in/Enter angle from *XY*-plane

enter: **60 <R>**

prompt: Toggle angle from/Enter angle in *XY*-plane from *X*-axis

enter: **50 <R>**

prompt: CAmera...

enter: **H <R>** then **X <R>**

9. Make viewport (d) active, use DVIEW and pick the complete model remembering to enter <R>

prompt: CAmera/TArget...

enter: **PO <R>** – points option

prompt: Enter target point

enter: **50,0,0 <R>** – note where cursor is positioned!

prompt: Enter camera point

enter: **50,0,100 <R>**

prompt: CAmera...

enter: **X <R>** to end sequence.

10. DVIEW the model in viewport (e) and enter:
    (a) **TW <R>** – twist option
    (b) **180 <R>** – new view twist
    (c) **X <R>**.

11. With viewport (f), DVIEW the model and enter:
    (a) **CL <R>** – clip option
    (b) **F <R>** – front clip
    (c) **50 <R>** – distance from target
    (d) **X <R>**

12. Now HIDE each viewport and the result should be Fig. 32.1 which has been plotted with HIDEPLOT ON.

13. SHADE in each viewport which should give some interesting results as the various options which were entered can be viewed easily. The TWist and CLip displays are especially interesting?

14. REGENALL to end this exercise. You may want to save the final drawing for future reference?

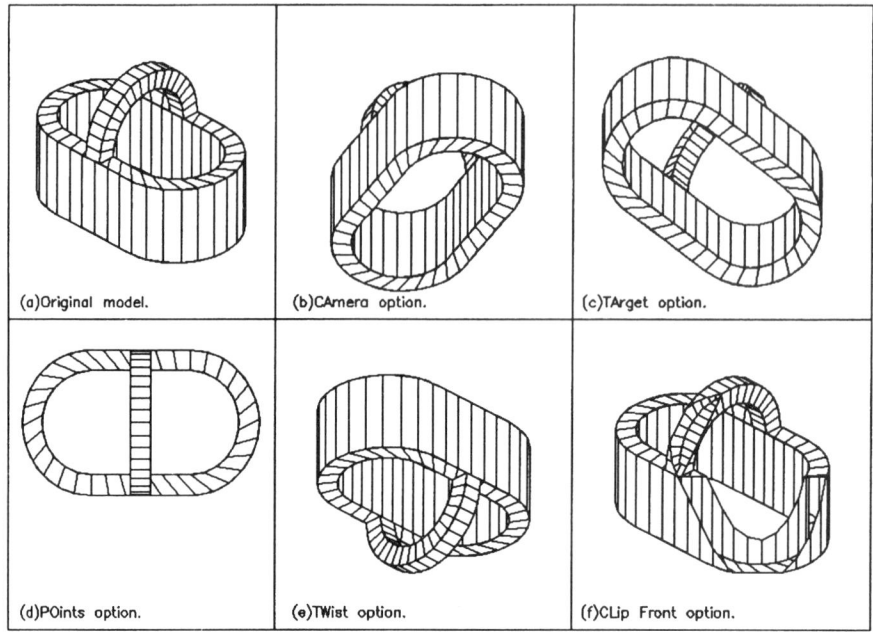

(a)Original model.   (b)CAmera option.   (c)TArget option.

(d)POints option.   (e)TWist option.   (f)CLip Front option.

Fig. 32.1 DVIEW exercise plotted with HIDEPLOT ON.

THE DVIEW OPTIONS

The various options of the DVIEW command are:

CAmera   turns your viewpoint as you look at the target.

TArget   turns the target viewpoint relative to the camera (you).

Distance alters the distance between the camera and the target. Also introduces 'perspective' to the model.

POints   allows the user to pick a point on the target at which to point the camera. The camera position is also entered.

PAn   pans between two points.

Zoom   zooms 'in' and 'out' of the model.

TWist   rotates the target about the XY plane.

CLip   removes parts of the target (model) the user entering the distance. The clip can be at the front or back of the target and the model is 'clipped' by a cutting plane which is perpendicular to the existing VPOINT plane.

Hide   hides the target.

Undo   undoes the last option entered. It can be used to undo all of the DVIEW options.

eXit   exits the DVIEW command. The model is re-configured with all the options which have been entered.

DVIEW EXERCISE 2

In this exercise we will investigate all the DVIEW options, one after the other. The sequences will be given as a series of keyboard entries and the prompt line will be omitted. There will be a brief note about each entry.

1. Open drawing **EX32_2** from the 3DPACK directory to display a two-viewport configuration of a 3D objects model. The right viewport should be active.
2. With the DVIEW command, select the complete model then enter the following sequence, continuing until the X is encountered:
   (a) CA – camera option
      60 – angle from XY-plane
      20 – angle in XY-plane from X-axis
      H – hide option
      U – undo hide effect
   (b) TA – target option
      30 – angle from XY-plane
      30 – angle in XY-plane from X-axis
      H – hide option
      U – undo hide effect
   (c) PO – points option
      80,80,80 – target point
      0,150,150 – camera point
      H – hide option
      U – undo hide effect
   (d) PA – pan option
      0,0,0 – displacement base point
      @0,100,0 – second point
   (e) Z – zoom option
      2 – zoom scale factor (drawing enlarged – don't worry)
   (f) D – distance option
      750 – new distance (note perspective of model and new icon)
   (g) TW – twist option
      –90 – new twist angle
      H – hide effect
      U – undo hide effect
   (h) CL – clip option
      F – front clip
      75 – clip distance
      B – back clip
      0 – clip distance
   (i) X – end sequence.
3. Now HIDE the model to 'see' the result of the DVIEW sequence. Figure 32.2 displays the model.
4. When complete the icon should have changed and it should be an oblong box. This is the perspective icon, and your model is displayed with true artistic perspective.
5. Enter the ZOOM command and the message:
   That command may not be invoked in a perspective view will be obtained.

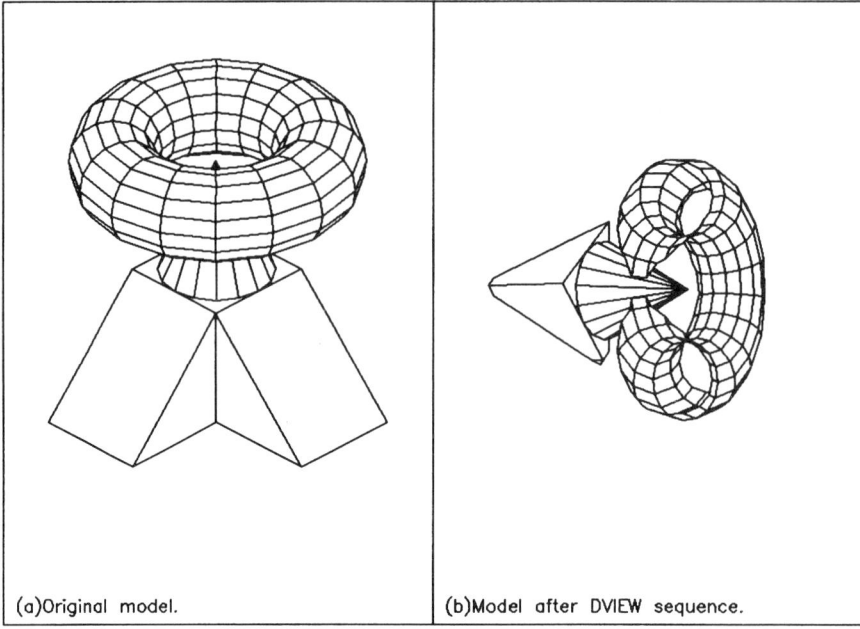

Fig. 32.2   DVIEW exercise 2 after the DVIEW option sequence.

SUMMARY

1. DVIEW allows models to be viewed in perspective.
2. The command is non-viewport dependent, i.e. modifications to the model in one viewport, is not reflecting in the other viewports.
3. The command has several options.
4. The camera and target options are very similar to the VPOINT Rotate command.
5. The camera and target options can give the same result, depending on the angles entered.
6. The distance option invokes perspective views.
7. Twist can be used to 'invert' a model.
8. The clip option will allow the user to 'see' into models.

ACTIVITY

Solve the riddle of the pyramid

This activity consists of a mystery for you to solve. A pyramid has been constructed and there is an entrance to a chamber inside the pyramid. There is a message on the chamber door which has to be deciphered, and it is your task to find out that this message is.

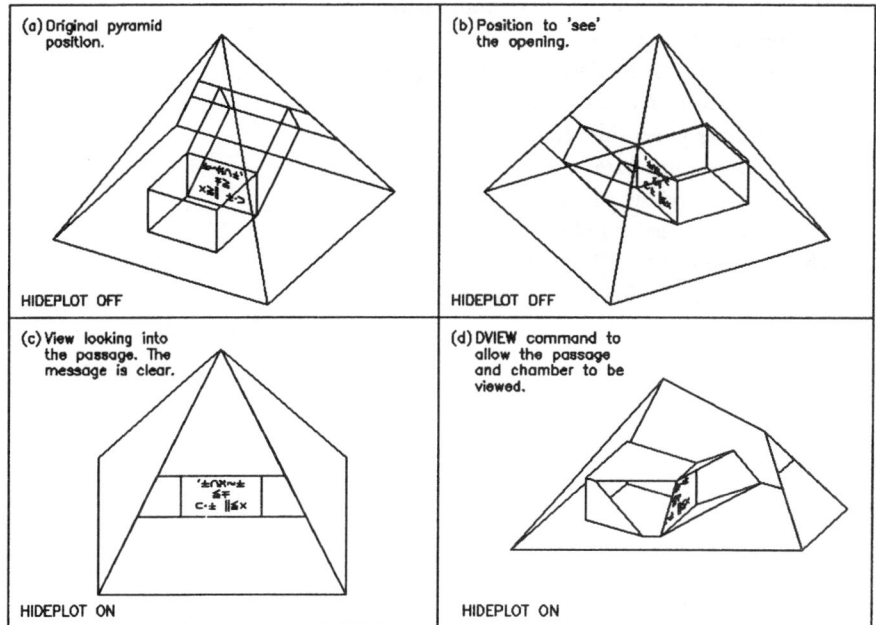

Fig. 32.3   Pyramid mystery solved?

Open **ACT_13** from the 3DPACK directory. The screen will display a four viewport configuration of the pyramid – the same view in each viewport. The pyramid was constructed from:

1. Four green-faced sides.
2. A blue face base.
3. A yellow passage made from four faced sides.
4. A cyan chamber made from a box.
5. A white message on the chamber door.

Your tasks are as follows:

1. Leave viewport (a) untouched.
2. Alter the viewpoint in viewport (b) so that the passage opening is being viewed from the front.
3. Alter the vpoint in viewport (c) to 'see' directly into the passage, i.e. you have to view the face it is on perpendicularly. A UCS has been created to make this easy!
4. Enter model space and zoom in on the passage opening. The message on the chamber door should be clear.
5. Unfortunately this message is in code. Investigate the text styles which have been created. One of them was used to write the message.
6. Having the correct style should enable you to decipher the message, but it will require you to obtain the alphabet for this style and work backwards.
7. In viewport (d) use the DVIEW command to clip away the pyramid so that the passageway and the chamber can be seen.
8. Figure 32.3 displays the four viewports as they should be at the end of your investigation.

9. SHADE each viewport and you should have:
   - (a) two green sloped sides.
   - (b) two green sloped sides and a yellow passage opening
   - (c) three green sloped sides and a cyan chamber front
   - (d) cut-away green sides (three off)
       cut-away blue base
       cut-away yellow passage
       cut-away cyan chamber.

10. The message is ??????????????????

11. If you cannot decipher it, please write!

33

Slides in 3D

Slides are 'snapshots' of a screen drawing. They cannot be edited once they are made, but they can be viewed. Several slides can be 'run together' to make a slide show, which is animation with AutoCAD without using additional add-on packages.

In this chapter we will view an animated slide show and then make several slides of our own for viewing.

EXERCISE 1 – AN ANIMATED SLIDE SHOW

1. Open drawing **EX33_1** from the 3DPACK directory.
2. The screen displays a single viewport with nothing in it.
3. From the screen menu select **UTILITY**
SCRIPT
   prompt: Select Script File dialogue box
   respond: (a)   pick SHOW – turns blue
            (b)   pick OK
4. The screen will display a cube revolving inside a torus. The torus is turning in the opposite direction from the cube. We have an animated 3D drawing!
5. When you are fed up with the animation, it can be cancelled with **CTRL C.**
6. The screen will display the slide at the moment the slide show was cancelled.
7. To return to the original drawing screen enter REDRAW.

VIEWING EXISTING SLIDES

1. From the screen menu select **UTILITY**
SLIDES
VSLIDE
   prompt: Select Slide File dialogue box
   respond: pick SLMOD13 then OK.
2. The screen will display the slide SLMOD13, which was one of the slides used in the slide show.
3. Try and erase the torus or the box – you cannot as they are not entities.
4. Enter REDRAW to remove the slide and return to the original screen.

MAKING AND VIEWING SLIDES

Slides of a screen display are very easy to make.

1. Open drawing **EX33_2** from the 3DPACK directory. The screen will display a four-viewport configuration of a 3D model created from 3D objects (one box and six cylinders). The cylinders have been coloured and the lower right viewport is active – Fig. 33.1.
2. SHADE each viewport to 'see' the arrangement of the model then REGENALL.
3. With the lower-right viewport active, SHADE the model and select **UTILITY**
                                                                                 SLIDES
                                                                                 MSLIDE

    prompt: Create Slide File dialogue box
    check: (a) file extension (pattern) is *.sld
              (b) directory name is C:\3DPACK
              (c) files – several SLMOD? names which were used in the first exercise
    enter: **SL1** at the file name box then pick OK.
4. Activate the lower left viewport, SHADE it then enter **MSLIDE** <R> at the command line and:

    prompt: Create Slide File dialogue box (SL1 added to list?)
    enter: **SL2** as file name then OK
5. With the top to viewports active, SHADE each then make slides of them with the names:

    top left: SL3
    top right: SL4

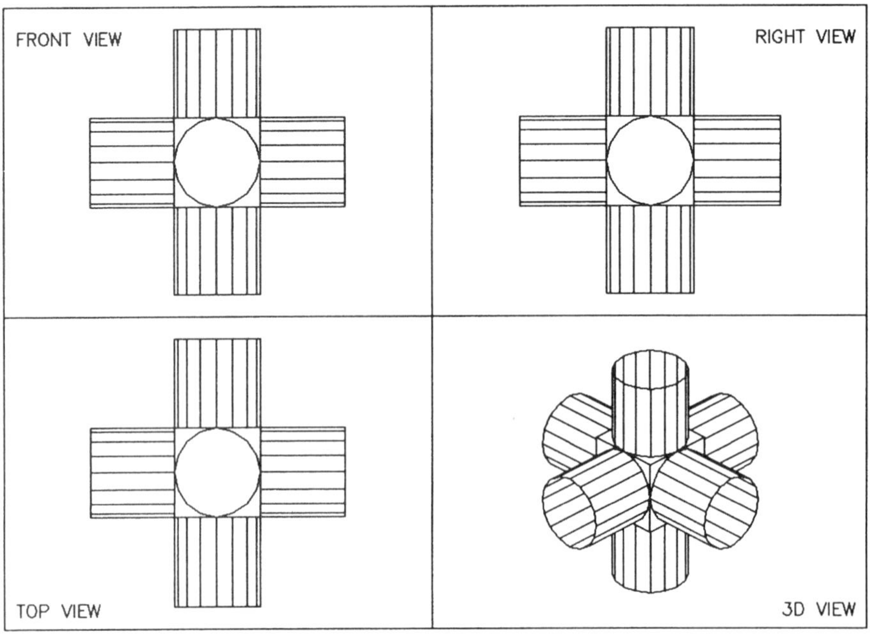

Fig. 33.1   Three-dimensional model for slide creation.

6. REGENALL to return screen to original drawing and make the lower right viewport active.
7. At the command line enter **VSLIDE** <R> and pick **SL1** from the file list then OK.
8. Slide SL1 will be displayed in the lower right viewport.
9. Repeat the VSLIDE command and view slides SL2, SL3 and SL4.
10. Enter REDRAW to return the original drawing.
11. Enter Paper space and view the slides SD–A, SD–B, SD–C and SD–D which have been created for you in the 3DPACK directory.
12. These slides are displayed 'full screen' size and also display the viewport name. The slides you created did not display the viewport name.
13. Redraw and return to Model space.

Question

1. Why did slides SD–A, etc. fill the screen and the slides SL1, etc. did not?
2. Why did slides SD–A, etc. have the viewport name and the slides SL1, etc. did not?

SLIDE SHOWS

A slide show is a sequence of slides which are 'run together'. The slides can have delays between them to allow the slides to be viewed on the screen for some time. If no delay is added between slides, then the slides appear on the screen one-after-the-other to give an 'animated' effect. Slide shows require that the user writes a SCRIPT text file with the extension .SCR, but it is not my intention to investigate these in this book.

SUMMARY

1. Slides are screen 'snapshots'.
2. Slides can be used for a variety of purposes:
   (a) animating a drawing
   (b) showing the 'build-up' of a 3D model
   (c) displaying different viewports of a model, etc.
3. There are only two slide commands – MSLIDE and VSLIDE.
4. Slides can be made into a slide show.
5. A slide file does not use a lot of memory when compared to a corresponding drawing file.

34

Three-dimensional rendering

AutoCAD has rendering capabilities. The render 'package' with R12 is not a complete render system, in that shadows cannot be added. It is possible to add lights and finishes to models which will greatly enhance the model appearance and impress both employers and customers. Rendered images can be extracted from AutoCAD in BMP format for colour laser printing.

Only certain types of models can be rendered with AutoCAD, these being:
- extruded models
- 3D faced models
- 3D ruled-surfaced models
- 3D revolved-surface models
- 3D objects
- solid models – not considered here.

It is not my intention to cover render as a topic in this chapter. My aim is to show the user the potential of the render facility within AutoCAD and this will be achieved by a series of exercises covering the different render 'types'. All of the examples are previous exercises from various chapters in the book and you should be familiar with them when you 'see' them on the screen.

RENDER EXAMPLE 1 – AN EXTRUDED MODEL

1. Open drawing **EX34_1** from the 3DPACK directory.
2. The screen displays a single viewport of an extruded model from section 2 – Fig. 34.1.
3. From the menu bar select **Render**
                       Render
4. The screen flips to Text with the following:
   Initializing...
   Initializing AVE Render
   Please configure RENDER
   AutoCAD's combined Display/Rendering driver supports < 128 colors
   prompt: Press RETURN to continue
   enter: **<RETURN>**   (1)
   prompt: Select rendering display device:

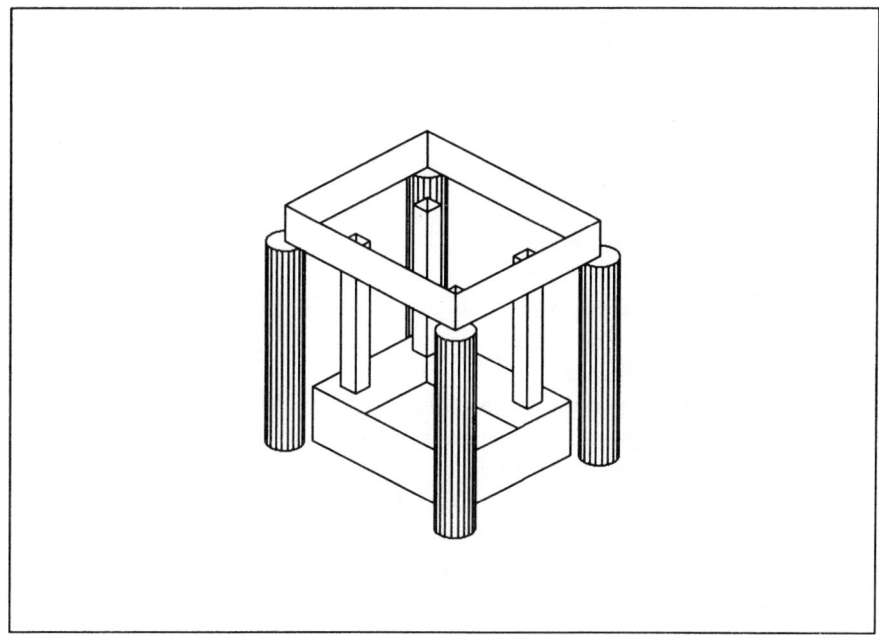

Fig. 34.1   Render example 1 – extruded drawing.

        1. AutoCAD's configured P386ADI combined disp/rend driver
        2. P386 Autodesk Device Interface rendering driver
        3. None (Null rendering device)

prompt: Rendering selection<1>
enter: **<RETURN>** (2)
prompt: Default <1> selected
then: Do you want to do detailed configuration of SVADT's rendering feature?<N>
enter: **<RETURN>** (3)
prompt: Select mode to run display/render combined driver
  1. Render to display viewport
  2. Render to rendering screen
then: AutoCAD driver does not support rendering to a viewport
Rendering screen automatically selected
Press <RETURN> to continue
enter: **<RETURN>** (4)
prompt: Select rendering hard copy device
        1. None (Null rendering device)
        2. P386 AutoDESK Device Interface rendering driver
        3. Rendering file (256 color map)
        4. Rendering file (continuous color)
then: Rendering hard copy selection<1>
enter: **<RETURN>** (5)

prompt: Default <1> selected
           Enabling handles... done
           Initializing preferences... done
           Using current view
           Default scene selected
           Projecting objects into view plane
           Applying parallel projection
           Calculate extents for faces
           Sorting 288 triangles by depth
           Checking 288 triangles for obscuration
           Calculate shading and assign colors
           Outputting triangles

5. The result of all this is that the screen converts to black and a rendered image of the extruded model is displayed. This image is in the colours of the model entities and is displayed with hidden line removal.
6. The image is rather 'bland and bright' but is the 'basic rendering' produced without any attempt at adding lights to the image.
7. Return to the drawing screen with <RETURN>

Note

1. The five <RETURN> keyboard presses are necessary when a new model is being rendered for the first time. At our level, the default selections are adequate.
2. If a drawing has been saved which was previously rendered, then the five <RETURNS> will not be necessary.

RENDER EXAMPLE 2 – A 3D-FACED MODEL

1. Open drawing **EX34_2** from the 3DPACK directory.
2. The screen displays a 3D faced hexagonal pyramid with coloured sides – Fig. 34.2.
3. When the drawing has been displayed, the command line displays:
   Initializing AVE Render, i.e. the 3d model was saved with Render 'loaded'.
4. Note the 'light' icon about mid-left screen.
5. From the screen menu select **RENDER**
                          RENDER
6. The result is a rendered image of the pyramid with the yellow face bright and the green and blue faces dull. A light is shining towards the yellow face.
7. Change the viewpoints to the following and RENDER the screen after each entry:
   (a) VPOINT Rotate 215,40 – yellow and blue bright, green dull
   (b) VPOINT Rotate 160, 20 – yellow and blue
   (c) VPOINT Rotate 160,–50 – no base on model?

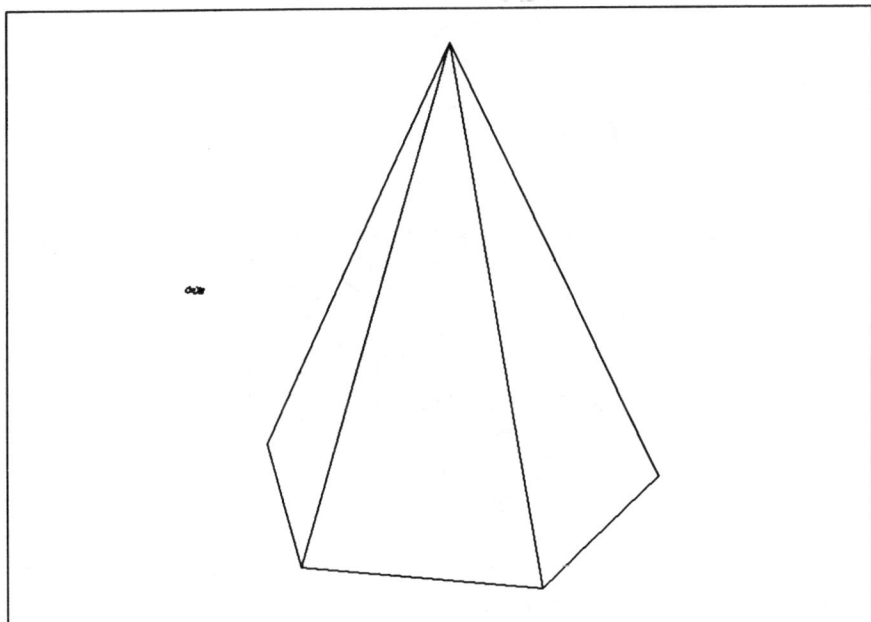

Fig. 34.2   Render example 2 – a FACED wire-frame model.

TASK

Can you add a coloured base to the pyramid?

RENDER EXAMPLE 3 – A 3D RULED-SURFACE MODEL

This model has had two lights added.
1. Open drawing **EX34_3** from the 3DPACK directory – Fig. 34.3.
2. Note the two light symbols.
3. Render the model. The lights are shining:
   (a) towards the left side of the model
   (b) towards the arch undersides.
4. Try some other viewpoints, e.g.
   (a) VPOINT Rotate 315,–30 – no base again?
   (b) VPOINT Rotate 30,30 – generally dull
   (c) VPOINT Rotate 210,50 – dull again?
   (d) VPOINT Rotate 290,–10.

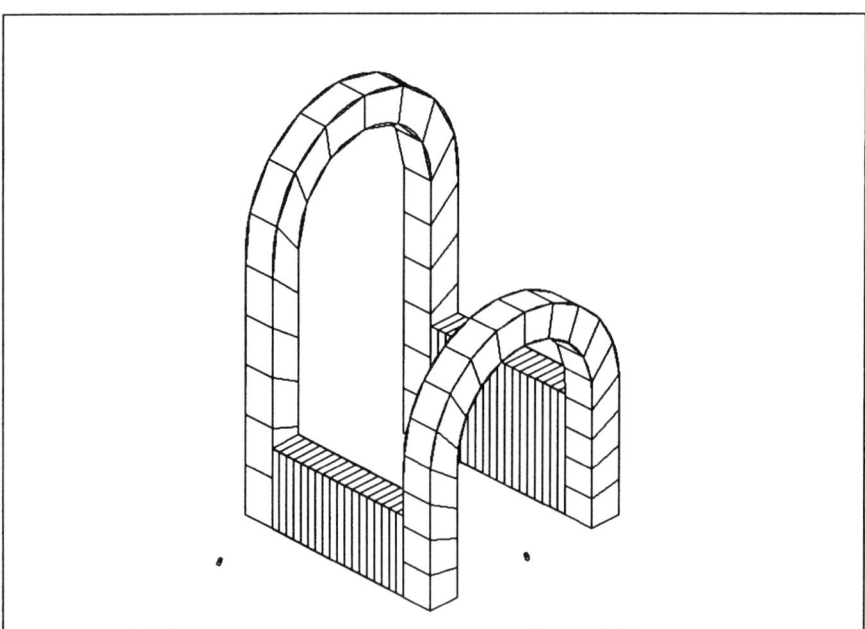

Fig. 34.3   Render example 3 – a RULED-SURFACE model.

RENDER EXAMPLE 4 – A REVOLVED-SURFACE 3D MODEL

AutoCAD render allows the user to added 'finishes' to a model which can result in either a dullish or shiny surface. This current example has had a finish added to the model.

1. Open drawing **EX34_4** from the 3DPACK directory.
2. The screen displays the 3D glass created from the Revolved surfaces section – Fig. 34.4.
3. Three lights have been added.
4. Render the model.
5. Change the viewpoint to:
   (a) VPOINT 1,0,0
   (b) VPOINT Rotate 80,80
   (c) VPOINT 1,1,1
   (d) VPOINT 0,1,0

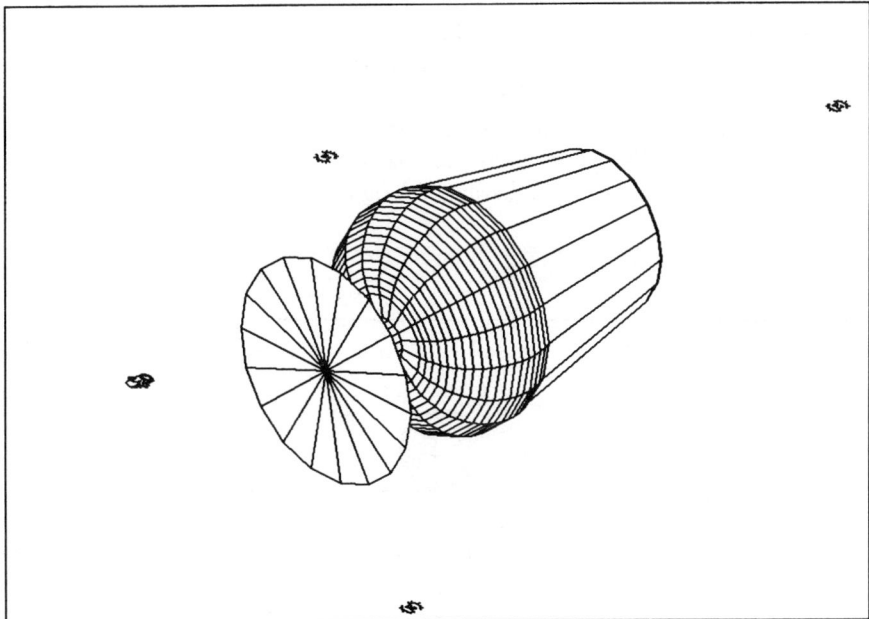

Fig. 34.4   Render example 4 – a REVSURF model.

RENDER EXAMPLE 5 – 3D OBJECTS

When lights are added to enable rendering, these lights are all 'on' at the one time. To allow the user to select different lights, the user creates 'scenes' of the model, adding the appropriate lights to the scenes as required. This example will investigate how scenes are rendered. It has had two lights added.

1. Open drawing **EX34_5** from the 3DPACK directory.
2. A 3D object model is displayed – Fig. 34.5.
3. From the menu bar select **Render**
   Scenes...
   prompt: Scenes dialogue box
   respond: **pick SCENE1 then OK**
4. From the screen menu select **RENDER:**
5. The model will be rendered, the image showing that the light is shining from the left side.
6. Return to the drawing screen.
7. From the screen menu select **SCENE...**
   SCENE2 from the dialogue box
   OK
8. The rendered image is lit from the right side.
9. Finally select SCENE3 from the Scenes dialogue box and render it.
10. This image is nice and bright, having both lights 'on' at the one time.

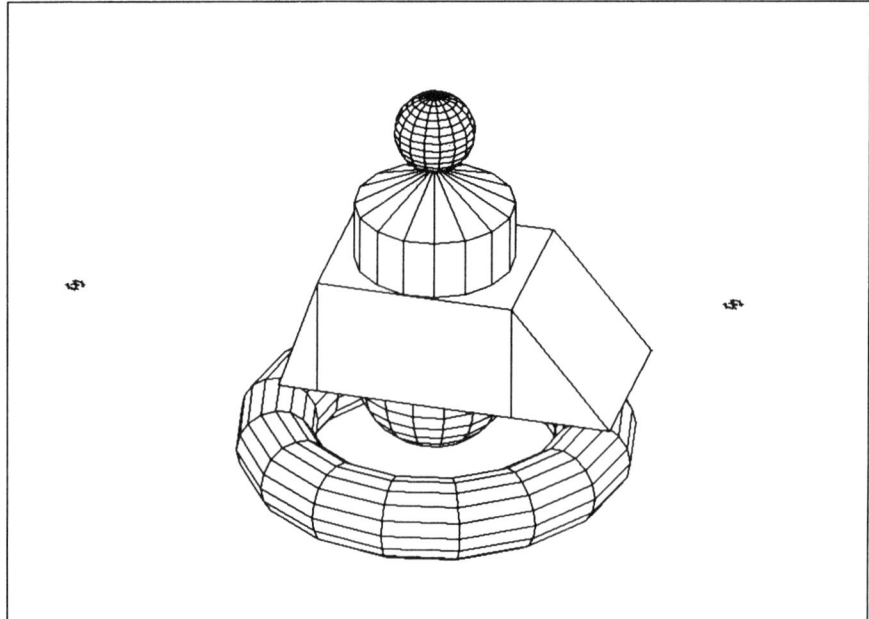

Fig. 34.5   Render example 5 – 3D objects.

AUTOCAD'S RENDERED DRAWINGS

Before leaving this section on rendering, it is worthwhile investigating two of AutoCAD's rendered drawings. When displayed, it will make you realise what can be achieved with rendering.

Note

The two drawings which will be rendered are stored in the TUTORIAL directory of ACAD12. It may be that your system does not have these drawings, in which case you will obviously not be able to 'see' the images. The drawings are supplied with the AutoCAD R12 pack.

1. From the menu bar select **File**

    > **Open** (discarding any changes)
    > \   – double left click
    > **ACAD12**   – or similar name
    > **TUTORIAL** – double left click
    > **KITCHEN2** – drawing name
    > **OK**

2. The screen displays one large viewport and three small viewports of a kitchen.
3. Ensure the large viewport is active and select from the menu bar **Render–Render**.
4. Nice rendered image?
5. Now open the drawing **PINS2** from the TUTORIAL directory and render the large viewport.

SUMMARY

Rendering allows coloured images to be displayed. The final effect is very pleasing and adds the 'final touch' to 3D models.

35

Conclusion

This package has introduced the user to the basic principles of 3D draughting. I have tried to make the exercises as interesting and varied as possible and I hope that you have enjoyed working through them. It is only with practice that you will become proficient at draughting in 3D, but you should now have the basic knowledge to proceed. I would summarize the most important aspects of 3D draughting as:

• the UCS icon – positioning, saving, recall
• coordinate input in 3D
• paper space viewports.

Once you are feel competent at 3D draughting, your next step is solid modelling, which is a fascinating topic, and much easier that you would imagine.

COMMENTS

Any comments you have about this package (good or bad) would be more than welcome.

Index